Machine Learning Engineering with MLflow

Manage the end-to-end machine learning life cycle with MLflow

Natu Lauchande

BIRMINGHAM—MUMBAI

Machine Learning Engineering with MLflow

Publishing Product Manager: Reshma Raman
Senior Editor: David Sugarman
Content Development Editor: Sean Lobo
Technical Editor: Manikandan Kurup
Copy Editor: Safis Editing
Project Coordinator: Aparna Ravikumar Nair
Proofreader: Safis Editing
Indexer: Pratik Shirodkar
Production Designer: Sinhayna Bais

First published: August 2021

Production reference: 1220721

Published by Packt Publishing Ltd.
Livery Place
35 Livery Street
Birmingham
B3 2PB, UK.

ISBN 978-1-80056-079-6

www.packt.com

Contributors

About the author

Natu Lauchande is a principal data engineer in the fintech space currently tackling problems at the intersection of machine learning, data engineering, and distributed systems. He has worked in diverse industries, including biomedical/pharma research, cloud, fintech, and e-commerce/mobile. Along the way, he had the opportunity to be granted a patent (as co-inventor) in distributed systems, publish in a top academic journal, and contribute to open source software. He has also been very active as a speaker at machine learning/tech conferences and meetups.

About the reviewer

Hitesh Hinduja is an ardent AI enthusiast working as a Senior Manager in AI at Ola Electric, where he leads a team of 20+ people in the areas of machine learning, deep learning, statistics, computer vision, natural language processing, and reinforcement learning. He has filed 14+ patents in India and the US and has numerous research publications under his name. Hitesh has been associated in research roles at India's top B-schools: Indian School of Business, Hyderabad, and the Indian Institute of Management, Ahmedabad. He is also actively involved in training and mentoring and has been invited as a guest speaker by various corporates and associations across the globe.

Table of Contents

Section 2: Model Development and Experimentation

3
Your Data Science Workbench

4
Experiment Management in MLflow

5
Managing Models with MLflow

Section 3: Machine Learning in Production

6
Introducing ML Systems Architecture

7
Data and Feature Management

11

Performance Monitoring

12

Advanced Topics with MLflow

Other Books You May Enjoy

Index

Preface

Implementing a product based on machine learning can be a laborious task. There is a general need to reduce the friction between different steps of the machine learning development life cycle and between the teams of data scientists and engineers that are involved in the process.

Machine learning practitioners such as data scientists and machine learning engineers operate with different systems, standards, and tools. While data scientists spend most of their time developing models in tools such as Jupyter Notebook, when running in production, the model is deployed in the context of a software application with an environment that's more demanding in terms of scale and reliability.

In this book, you will be introduced to MLflow and machine learning engineering practices that will aid your machine learning life cycle, exploring data acquisition, preparation, training, and deployment. The book's content is based on an open interface design and will work with any language or platform. You will also gain benefits when it comes to scalability and reproducibility.

By the end of this book, you will be able to comfortably deal with setting up a development environment for models using MLflow, framing your machine learning problem, and using a standardized framework to set up your own machine learning systems. This book is also particularly handy if you are implementing your first machine learning project in production.

Who this book is for

This book is geared toward software, machine learning, and data science professionals or enthusiasts who want to explore the engineering side of machine learning systems in production. Machine learning practitioners will be able to put their knowledge to work with this practical guide to MLflow. The book takes a hands-on approach to implementation and associated methodologies that will have you up and running with MLflow in no time. The basic requirements for this book are experience in Python programming and knowledge of the Bash terminal and commands.

What this book covers

Chapter 1, *Introducing MLflow*, will be an overview of the different features of MLflow, guiding you in installing and exploring the core features of the platform. After reading this chapter, you will be able to install and operate your MLflow environment locally.

Chapter 2, *Your Machine Learning Project*, introduces the focus of the book. The approach of this book is to work through a practical business case, namely stock market prediction, and, through this use case, explore all the different features of MLflow. A problem-framing framework will be used to get you deeply familiar with the example used in the book. A sample pipeline will be created for use in the remainder of the book.

Chapter 3, *Your Data Science Workbench*, helps you understand how to use MLflow to create your local environment so that you can develop your machine learning projects locally using all the different features provided by MLflow.

Chapter 4, *Experiment Management in MLflow*, is where you will gain practical experience of stock prediction by creating different models and comparing the metrics of different runs in MLflow. You will be guided as to how to deploy a tracking server so that many machine learning practitioners can share metrics and improve a model.

Chapter 5, *Managing Models with MLflow*, looks at the different features for model creation in MLflow. Built-in models, such as PyTorch and TensorFlow models, will be covered alongside custom models not available in MLflow. A model life cycle will be introduced alongside the Model Registry feature of MLflow.

Chapter 6, *Introducing ML Systems Architecture*, talks about the need to architect machine learning systems properly and how MLflow fits in the picture of an end-to-end machine learning system.

Chapter 7, *Data and Feature Management*, introduces data and feature management. The importance of feature generation will be made clear, as will how to use feature streams to log model results with MLflow.

Chapter 8, *Training Models with MLflow*, is where the complete training pipeline infrastructure will be described and developed for the problem at hand, with the use of MLflow-specific features.

Chapter 9, *Deployment and Inference with MLflow*, is where an end-to-end deployment infrastructure for our machine learning system, including the inference component, will be exposed using the API and batch features of MLflow. The cloud-enabled features of MLflow will also be described.

Chapter 10, Scaling Up Your Machine Learning Workflow, covers integration with high-performance/big data libraries that allow MLflow systems to scale for large volumes of data.

Chapter 11, Performance Monitoring, explores the important area of machine learning operations and how to ensure a smooth ride for the production systems developed in the book using best practices and operational patterns.

Chapter 12, Advanced Topics with MLFlow, presents advanced case studies with complete MLflow pipelines. The case studies use different types of models from the ones looked at in the rest of the book to ensure a breadth of feature coverage for MLflow.

To get the most out of this book

Ideally, before getting started with the book, you should have a good grasp of the Python programming language and should have already created basic machine learning models. One introductory course in machine learning will help contextualize the concepts discussed in the book.

Software/hardware covered in the book	Operating system requirements
MLflow 1.18	Windows, macOS, or Linux
Python 3.7+	

If you are using the digital version of this book, we advise you to type the code yourself or access the code from the book's GitHub repository (a link is available in the next section). Doing so will help you avoid any potential errors related to the copying and pasting of code.

Download the example code files

You can download the example code files for this book from GitHub at `https://github.com/PacktPublishing/Machine-Learning-Engineering-with-MLflow`. If there's an update to the code, it will be updated in the GitHub repository.

We also have other code bundles from our rich catalog of books and videos available at `https://github.com/PacktPublishing/`. Check them out!

Download the color images

We also provide a PDF file that has color images of the screenshots and diagrams used in this book. You can download it here: `https://static.packt-cdn.com/downloads/9781800560796_ColorImages.pdf`.

Conventions used

There are a number of text conventions used throughout this book.

Code in text: Indicates code words in text, database table names, folder names, filenames, file extensions, pathnames, dummy URLs, user input, and Twitter handles. Here is an example: "The `model.pkl` file contains a serialized version of the model"

A block of code is set as follows:

```
import mlflow
from sklearn.linear_model import LogisticRegression
mlflow.sklearn.autolog()
with mlflow.start_run():
    clf = LogisticRegression()
    clf.fit(X_train, y_train)
```

Any command-line input or output is written as follows:

```
docker build -t stockpred -f dockerfile
```

Bold: Indicates a new term, an important word, or words that you see onscreen. For instance, words in menus or dialog boxes appear in bold. Here is an example: "**MLflow** is an open source platform for the **machine learning (ML)** life cycle"

> Tips or important notes
> Appear like this.

Get in touch

Feedback from our readers is always welcome.

General feedback: If you have questions about any aspect of this book, email us at customercare@packtpub.com and mention the book title in the subject of your message.

Errata: Although we have taken every care to ensure the accuracy of our content, mistakes do happen. If you have found a mistake in this book, we would be grateful if you would report this to us. Please visit www.packtpub.com/support/errata and fill in the form.

Piracy: If you come across any illegal copies of our works in any form on the internet, we would be grateful if you would provide us with the location address or website name. Please contact us at copyright@packt.com with a link to the material.

If you are interested in becoming an author: If there is a topic that you have expertise in and you are interested in either writing or contributing to a book, please visit authors.packtpub.com.

Share Your Thoughts

Once you've read *Machine Learning Engineering with MLflow*, we'd love to hear your thoughts! Scan the QR code below to go straight to the Amazon review page for this book and share your feedback.

https://packt.link/r/1-800-56079-6

Your review is important to us and the tech community and will help us make sure we're delivering excellent quality content.

Section 1: Problem Framing and Introductions

This section will introduce you to a framework for stating machine learning problems in a concise and clear manner using MLflow.

The following chapters are covered in this section:

1
Introducing MLflow

MLflow is an open source platform for the **machine learning (ML)** life cycle, with a focus on *reproducibility*, *training*, and *deployment*. It is based on an open interface design and is able to work with any language or platform, with clients in Python and Java, and is accessible through a REST API. Scalability is also an important benefit that an ML developer can leverage with MLflow.

In this chapter of the book, we will take a look at how MLflow works, with the help of examples and sample code. This will build the necessary foundation for the rest of the book in order to use the concept to engineer an end-to-end ML project.

Specifically, we will look at the following sections in this chapter:

- What is MLflow?
- Getting started with MLflow
- Exploring MLflow modules

Technical requirements

For this chapter, you will need the following prerequisites:

- The latest version of Docker installed in your machine. In case you don't have the latest version, please follow the instructions at the following URL: https://docs.docker.com/get-docker/.

- Access to a bash terminal (Linux or Windows).

- Access to a browser.

- Python 3.5+ installed.

- PIP installed.

What is MLflow?

Implementing a product based on ML can be a laborious task. There is a general need to reduce the friction between different steps of the ML development life cycle, and between teams of data scientists and engineers that are involved in the process.

ML practitioners, such as data scientists and ML engineers, operate with different systems, standards, and tools. While data scientists spend most of their time developing models in tools such as Jupyter Notebooks, when running in production, the model is deployed in the context of a software application with an environment that is more demanding in terms of scale and reliability.

A common occurrence in ML projects is to have the models reimplemented by an engineering team, creating a custom-made system to serve the specific model. A set of challenges are common with teams that follow bespoke approaches regarding model development:

- ML projects that run over budget due to the need to create bespoke software infrastructure to develop and serve models

- Translation errors when reimplementing the models produced by data scientists

- Scalability issues when serving predictions

- Friction in terms of reproducing training processes between data scientists due to a lack of standard environments

Companies leveraging ML tend to create their own (often extremely laborious) internal systems in order to ensure a smooth and structured process of ML development. Widely documented ML platforms include systems such as Michelangelo and FBLearner, from Uber and Facebook, respectively.

It is in the context of the increasing adoption of ML that MLflow was initially created at Databricks and open sourced as a platform, to aid in the implementation of ML systems.

MLflow enables an everyday practitioner in one platform to manage the ML life cycle, from iteration on model development up to deployment in a reliable and scalable environment that is compatible with modern software system requirements.

Getting started with MLflow

Next, we will install MLflow on your machine and prepare it for use in this chapter. You will have two options when it comes to installing MLflow. The first option is through a Docker container-based recipe provided in the repository of the book: `https://github.com/PacktPublishing/Machine-Learning-Engineering-with-Mlflow.git`.

To install it, follow these instructions:

1. Use the following commands to install the software:

   ```
   $ git clone https://github.com/PacktPublishing/Machine-Learning-Engineering-with-Mlflow.git

   $ cd Machine-Learning-Engineering-with-Mlflow

   $ cd Chapter01
   ```

2. The Docker image is very simple at this stage: it simply contains MLflow and sklearn, the main tools to be used in this chapter of the book. For illustrative purposes, you can look at the content of the `Dockerfile`:

   ```
   FROM jupyter/scipy-notebook
   RUN pip install mlflow
   RUN pip install sklearn
   ```

3. To build the image, you should now run the following command:

   ```
   docker build -t chapter_1_homlflow
   ```

4. Right after building the image, you can run the `./run.sh` command:

   ```
   ./run.sh
   ```

   > **Important note**
   > It is important to ensure that you have the latest version of Docker installed on your machine.

5. Open your browser to `http://localhost:888` and you should be able to navigate to the `Chapter01` folder.

In the following section, we will be developing our first model with MLflow in the Jupyter environment created in the previous set of steps.

Developing your first model with MLflow

From the point of view of simplicity, in this section, we will use the built-in sample datasets in sklearn, the ML library that we will use initially to explore MLflow features. For this section, we will choose the famous `Iris` dataset to train a multi-class classifier using MLflow.

The Iris dataset (one of sklearn's built-in datasets available from `https://scikit-learn.org/stable/datasets/toy_dataset.html`) contains the following elements as features: sepal length, sepal width, petal length, and petal width. The target variable is the class of the iris: Iris Setosa, Iris Versocoulor, or Iris Virginica:

1. Load the sample dataset:

    ```
    from sklearn import datasets
    from sklearn.model_selection import train_test_split
    dataset = datasets.load_iris()
    X_train, X_test, y_train, y_test = train_test_
    split(dataset.data, dataset.target, test_size=0.4)
    ```

2. Next, let's train your model.

    Training a simple machine model with a framework such as scikit-learn involves instantiating an estimator such as `LogisticRegression` and calling the `fit` command to execute training over the `Iris` dataset built in scikit-learn:

    ```
    from sklearn.linear_model import LogisticRegression
    clf = LogisticRegression()
    clf.fit(X_train, y_train)
    ```

    The preceding lines of code are just a small portion of the ML **Engineering** process. As will be demonstrated, a non-trivial amount of code needs to be created in order to productionize and make sure that the preceding training code is usable and reliable. One of the main objectives of MLflow is to aid in the process of setting up ML systems and projects. In the following sections, we will demonstrate how MLflow can be used to make your solutions robust and reliable.

3. Then, we will add MLflow.

With a few more lines of code, you should be able to start your first MLflow interaction. In the following code listing, we start by importing the `mlflow` module, followed by the `LogisticRegression` class in scikit-learn. You can use the accompanying Jupyter notebook to run the next section:

```
import mlflow
from sklearn.linear_model import LogisticRegression
mlflow.sklearn.autolog()
with mlflow.start_run():
    clf = LogisticRegression()
    clf.fit(X_train, y_train)
```

The `mlflow.sklearn.autolog()` instruction enables you to automatically log the experiment in the local directory. It captures the metrics produced by the underlying ML library in use. **MLflow Tracking** is the module responsible for handling metrics and logs. By default, the metadata of an MLflow run is stored in the local filesystem.

4. If you run the following excerpt on the accompanying notebook's root document, you should now have the following files in your home directory as a result of running the following command:

```
$ ls -l
total 24
-rw-r--r-- 1 jovyan users 12970 Oct 14 16:30 chapther_01_
introducing_ml_flow.ipynb
-rw-r--r-- 1 jovyan users    53 Sep 30 20:41 Dockerfile
drwxr-xr-x 4 jovyan users   128 Oct 14 16:32 mlruns
-rwxr-xr-x 1 jovyan users    97 Oct 14 13:20 run.sh
```

The `mlruns` folder is generated alongside your notebook folder and contains all the experiments executed by your code in the current context.

The `mlruns` folder will contain a folder with a sequential number identifying your experiment. The outline of the folder will appear as follows:

```
├── 46dc6db17fb5471a9a23d45407da680f
│   ├── artifacts
│   │   └── model
│   │       ├── MLmodel
│   │       ├── conda.yaml
```

```
|    |              ├── input_example.json
|    |              └── model.pkl
|    ├── meta.yaml
|    ├── metrics
|    |    └── training_score
|    ├── params
|    |    ├── C
|    |    .....
|    └── tags
|         ├── mlflow.source.type
|         └── mlflow.user
└── meta.yaml
```

So, with very little effort, we have a lot of traceability available to us, and a good foundation to improve upon.

Your experiment is identified as UUID on the preceding sample by 46dc6db17fb5471a9a23d45407da680f. At the root of the directory, you have a yaml file named meta.yaml, which contains the content:

```
artifact_uri: file:///home/jovyan/
mlruns/0/518d3162be7347298abe4c88567ca3e7/artifacts
end_time: 1602693152677
entry_point_name: ''
experiment_id: '0'
lifecycle_stage: active
name: ''
run_id: 518d3162be7347298abe4c88567ca3e7
run_uuid: 518d3162be7347298abe4c88567ca3e7
source_name: ''
source_type: 4
source_version: ''
start_time: 1602693152313
status: 3
tags: []
user_id: jovyan
```

This is the basic metadata of your experiment, with information including start time, end time, identification of the run (`run_id` and `run_uuid`), an assumption of the life cycle stage, and the user who executed the experiment. The settings are basically based on a default run, but provide valuable and readable information regarding your experiment:

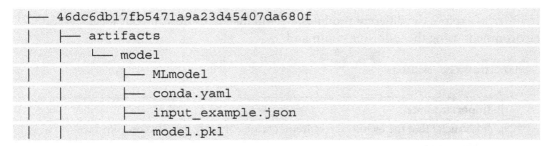

```
├── 46dc6db17fb5471a9a23d45407da680f
│   ├── artifacts
│   │   └── model
│   │       ├── MLmodel
│   │       ├── conda.yaml
│   │       ├── input_example.json
│   │       └── model.pkl
```

The `model.pkl` file contains a serialized version of the model. For a scikit-learn model, there is a binary version of the Python code of the model. Upon autologging, the metrics are leveraged from the underlying machine library in use. The default packaging strategy was based on a `conda.yaml` file, with the right dependencies to be able to serialize the model.

The `MLmodel` file is the main definition of the project from an MLflow project with information related to how to run inference on the current model.

The `metrics` folder contains the training score value of this particular run of the training process, which can be used to benchmark the model with further model improvements down the line.

The `params` folder on the first listing of folders contains the default parameters of the logistic regression model, with the different default possibilities listed transparently and stored automatically.

Exploring MLflow modules

MLflow modules are software components that deliver the core features that aid in the different phases of the ML life cycle. MLflow features are delivered through modules, extensible components that organize related features in the platform.

The following are the built-in modules in MLflow:

- **MLflow Tracking**: Provides a mechanism and UI to handle metrics and artifacts generated by ML executions (training and inference)
- **Mlflow Projects**: A package format to standardize ML projects

- **Mlflow Models**: A mechanism that deploys to different types of environments, both on-premises and in the cloud
- **Mlflow Model Registry**: A module that handles the management of models in MLflow and its life cycle, including state

In order to explore the different modules, we will install MLflow in your local environment using the following command:

```
pip install mlflow
```

> **Important note**
> It is crucial that the technical requirements are correctly installed on your local machine to allow you to follow along. You can also use the `pip` command with the required permissions.

Exploring MLflow projects

An MLflow project represents the basic unit of organization of ML projects. There are three different environments supported by MLflow projects: the Conda environment, Docker, and the local system.

> **Important note**
> Model details of the different parameters available on an MLProject file can be consulted in the official documentation available at `https://www.mlflow.org/docs/latest/projects.html#running-projects`.

The following is an example of an `MLproject` file of a `conda` environment:

```
name: condapred
conda_env:
  image: conda.yaml
entry_points:
  main:
    command: "python mljob.py"
```

In the `conda` option, the assumption is that there is a `conda.yaml` file with the required dependencies. MLflow, when asked to run the project, will start the environment with the specified dependencies.

The system-based environment will look like the following; it's actually quite simple:

```
name: syspred
entry_points:
  main:
    command: "python mljob.py"
```

The preceding system variant will basically rely on the local environment dependencies, assuming that the underlying operating system contains all the dependencies. This approach is particularly prone to library conflicts with the underlying operating system; it might be valuable in contexts where there is already an existing operating system environment that fits the project.

The following is a Docker environment-based MLproject file:

```
name: syspred
docker_env:
  image: stockpred-docker
entry_points:
  main:
    command: "python mljob.py"
```

Once you have your environment, the main file that defines how your project should look is the MLProject file. This file is used by MLflow to understand how it should run your project.

Developing your first end-to-end pipeline in MLflow

We will prototype a simple stock prediction project in this section with MLflow and will document the different files and phases of the solution. You will develop it in your local system using the MLflow and Docker installed locally.

> **Important note**
> In this section, we are assuming that MLflow and Docker are installed locally, as the steps in this section will be executed in your local environment.

The task in this illustrative project is to create a basic MLflow project and produce a working baseline ML model to predict, based on market signals over a certain number of days, whether the stock market will go up or down.

In this section, we will use a Yahoo Finance dataset available for quoting the BTC-USD pair in `https://finance.yahoo.com/quote/BTC-USD/` over a period of 3 months. We will train a model to predict whether the quote will be going up or not on a given day. A REST API will be made available for predictions through MLflow.

We will illustrate, step by step, the creation of an MLflow project to train a classifier on stock data, using the Yahoo API for financial information retrieved using the package's pandas data reader:

1. Add your `MLProject` file:

```
name: stockpred
docker_env:
  image: stockpred-docker
entry_points:
  main:
    command: "python train.py"
```

The preceding `MLProject` file specifies that dependencies will be managed in Docker with a specific image name. MLflow will try to pull the image using the version of Docker installed on your system. If it doesn't find it, it will try to retrieve it from Docker Hub. For the goals of this chapter, it is completely fine to have MLflow running on your local machine.

The second configuration that we add to our project is the main entry point command. The command to be executed will invoke in the Docker environment the `train.py` Python file, which contains the code of our project.

2. Add a Docker file to the project.

Additionally, you can specify the Docker registry URL of your image. The advantage of running Docker is that your project is not bound to the Python language, as we will see in the advanced section of this book. The MLflow API is available in a Rest interface alongside the official clients: Python, Java, and R:

```
FROM continuumio/miniconda:4.5.4

RUN pip install mlflow==1.11.0 \
    && pip install numpy==1.14.3 \
    && pip install scipy \
    && pip install pandas==0.22.0 \
    && pip install scikit-learn==0.20.4 \
```

```
        && pip install cloudpickle \
        && pip install pandas_datareader>=0.8.0
```

The preceding Docker image file is based on the open source package Miniconda, a free minimal installer with a minimal set of packages for data science that allow us to control the details of the packages that we need in our environment.

We will specify the version of MLflow (our ML platform), `numpy`, and `scipy` for numerical calculations. `Cloudpickle` allows us to easily serialize objects. We will use `pandas` to manage data frames, and `pandas_datareader` to allow us to easily retrieve the data from public sources.

3. Import the packages required for the project.

    On the following listing, we explicitly import all the libraries that we will use during the execution of the training script: the library to read the data, and the different `sklearn` modules related to the chosen initial ML model:

    ```
    import numpy as np
    import datetime
    import pandas_datareader.data as web
    from sklearn.model_selection import train_test_split
    from sklearn.ensemble import RandomForestClassifier
    from sklearn.metrics import classification_report
    from sklearn.metrics import precision_score
    from sklearn.metrics import recall_score
    from sklearn.metrics import f1_score
    import mlflow.sklearn
    ```

    We explicitly chose for the stock market movement detection problem a `RandomForestClassifier`, due to the fact that it's an extremely versatile and widely accepted baseline model for classification problems.

4. Acquire your training data.

    The component of the code that acquires the Yahoo Finance stock dataset is intentionally small, so we choose a specific interval of 3 months to train our classifier.

    The `acquire_training_data` method returns a `pandas` data frame with the relevant dataset:

    ```
    def acquire_training_data():
    ```

```
start = datetime.datetime(2019, 7, 1)
end = datetime.datetime(2019, 9, 30)
df = web.DataReader("BTC-USD", 'yahoo', start, end)
return df
```

The format of the data acquired is the classic format for financial securities in exchange APIs. For every day of the period, we retrieve the following data: the highest value of the stock, the lowest, opening, and close values of the stock, as well as the volume. The final column represents the adjusted close value, the value after dividends, and splits:

Date	High	Low	Open	Close	Volume	Adj Close
2019-09-26	8304.219727	7888.500000	8074.990234	8201.360352	390323710	8201.360352
2019-09-27	8345.030273	8044.450195	8201.360352	8223.650391	244186983	8223.650391
2019-09-28	8265.990234	7943.330078	8223.650391	8065.259766	226712656	8065.259766
2019-09-29	8372.240234	7736.959961	8065.259766	8314.620117	407250929	8314.620117
2019-09-30	8537.599609	8212.009766	8314.620117	8326.639648	355446344	8326.639648

Figure 1.1 – Excerpt from the acquired data

Figure 1.2 is illustrative of the target variable that we would like to achieve by means of the current data preparation process:

Date	High	Low	Open	Close	Volume	Adj Close	Delta	to_predict
2019-09-26	8304.219727	7888.500000	8074.990234	8201.360352	390323710	8201.360352	126.370117	1
2019-09-27	8345.030273	8044.450195	8201.360352	8223.650391	244186983	8223.650391	22.290039	1
2019-09-28	8265.990234	7943.330078	8223.650391	8065.259766	226712656	8065.259766	-158.390625	0
2019-09-29	8372.240234	7736.959961	8065.259766	8314.620117	407250929	8314.620117	249.360352	1
2019-09-30	8537.599609	8212.009766	8314.620117	8326.639648	355446344	8326.639648	12.019531	1

Figure 1.2 – Excerpt from the acquired data with the prediction column

5. Make the data usable by scikit-learn.

The data acquired in the preceding step is clearly not directly usable by `RandomForestAlgorithm`, which thrives on categorical features. In order to facilitate the execution of this, we will transform the raw data into a feature vector using the rolling window technique.

Basically, the feature vector for each day becomes the deltas between the current and previous window days. In this case, we use the previous day's market movement (1 for a stock going up, 0 otherwise):

```python
def digitize(n):
    if n > 0:
        return 1
    return 0


def rolling_window(a, window):
    """
        Takes np.array 'a' and size 'window' as
parameters
        Outputs an np.array with all the ordered
sequences of values of 'a' of size 'window'
        e.g. Input: ( np.array([1, 2, 3, 4, 5, 6]), 4 )
            Output:
                    array([[1, 2, 3, 4],
                           [2, 3, 4, 5],
                           [3, 4, 5, 6]])
    """
    shape = a.shape[:-1] + (a.shape[-1] - window + 1,
window)
    strides = a.strides + (a.strides[-1],)
    return np.lib.stride_tricks.as_strided(a,
shape=shape, strides=strides)


def prepare_training_data(data):
    data['Delta'] = data['Close'] - data['Open']
    data['to_predict'] = data['Delta'].apply(lambda d:
digitize(d))
    return data
```

The following example is illustrative of the data frame output produced with the binarized ups and downs of the previous days:

```
[[0. 1. 1. ... 1. 0. 0.]
 [1. 1. 0. ... 0. 0. 1.]
 [1. 0. 0. ... 0. 1. 0.]
 ...
 [0. 0. 0. ... 0. 1. 1.]
 [0. 0. 0. ... 1. 1. 0.]
 [0. 0. 1. ... 1. 0. 1.]]

Shape: (79, 14)
```

Figure 1.3 – Feature vector with binarized market ups and downs

6. Train and store your model in MLflow.

    This portion of the following code listing calls the data preparation methods declared previously and executes the prediction process.

    The main execution also explicitly logs the ML model trained in the current execution in the MLflow environment.

```
if __name__ == "__main__":
    with mlflow.start_run():
    training_data = acquire_training_data()
    prepared_training_data_df = prepare_training_
data(training_data)

    btc_mat = prepared_training_data_df.as_matrix()

    WINDOW_SIZE = 14

    X = rolling_window(btc_mat[:, 7], WINDOW_SIZE)[:-1,
:]
    Y = prepared_training_data_df['to_predict'].as_
matrix()[WINDOW_SIZE:]

    X_train, X_test, y_train, y_test = train_test_
split(X, Y, test_size=0.25, random_state=4284,
stratify=Y)

    clf = RandomForestClassifier(bootstrap=True,
```

```
criterion='gini', min_samples_split=2, min_weight_
fraction_leaf=0.0, n_estimators=50, random_state=4284,
verbose=0)

    clf.fit(X_train, y_train)

    predicted = clf.predict(X_test)

    mlflow.sklearn.log_model(clf, "model_random_forest")
    mlflow.log_metric("precision_label_0", precision_
score(y_test, predicted, pos_label=0))
    mlflow.log_metric("recall_label_0", recall_score(y_
test, predicted, pos_label=0))
    mlflow.log_metric("f1score_label_0", f1_score(y_test,
predicted, pos_label=0))
    mlflow.log_metric("precision_label_1", precision_
score(y_test, predicted, pos_label=1))
    mlflow.log_metric("recall_label_1", recall_score(y_
test, predicted, pos_label=1))
    mlflow.log_metric("f1score_label_1", f1_score(y_test,
predicted, pos_label=1))
```

The `mlflow.sklearn.log_model(clf, "model_random_forest")` method takes care of persisting the model upon training. In contrast to the previous example, we are explicitly asking MLflow to log the model and the metrics that we find relevant. This flexibility in the items to log allows one program to log multiple models into MLflow.

In the end, your project layout should look like the following, based on the files created previously:

```
├── Dockerfile
├── MLproject
├── README.md
└── train.py
```

7. Build your project's Docker image.

    In order to build your Docker image, you should run the following command:

    ```
    docker build -t stockpred -f dockerfile
    ```

This will build the image specified previously with the `stockpred` tag. This image will be usable in MLflow in the subsequent steps as the model is now logged into your local registry.

Following execution of this command, you should expect a successful Docker build:

```
---> 268cb080fed2
Successfully built 268cb080fed2
Successfully tagged stockpred:latest
```

8. Run your project.

In order to run your project, you can now run the MLflow project:

```
mlflow run .
```

Your output should look similar to the excerpt presented here:

```
MLFLOW_EXPERIMENT_ID=0 stockpred:3451a1f python train.py'
in run with ID '442275f18d354564b6259a0188a12575' ===
               precision    recall  f1-score   support

           0       0.61      1.00      0.76        11
           1       1.00      0.22      0.36         9

    accuracy                           0.65        20
   macro avg       0.81      0.61      0.56        20
weighted avg       0.79      0.65      0.58        20


2020/10/15 19:19:39 INFO mlflow.projects: === Run (ID
'442275f18d354564b6259a0188a12575') succeeded ===
```

This contains a printout of your model, the ID of your experiment, and the metrics captured during the current run.

At this stage, you have a simple, reproducible baseline of a stock predictor pipeline using MLflow that you can improve on and easily share with others.

Re-running experiments

Another extremely useful feature of MLflow is the ability to re-run a specific experiment with the same parameters as it was run with originally.

For instance, you should be able to run your previous project by specifying the GitHub URL of the project:

```
mlflow run https://github.com/PacktPublishing/Machine-Learning-
Engineering-with-MLflow/tree/master/Chapter01/stockpred
```

Basically, what happens with the previous command is that MLflow clones the repository to a temporary directory and executes it, according to the recipe on MLProject.

The ID of the experiment (or the name) allows you to run the project with the original parameters, thereby enabling complete reproducibility of the project.

The MLflow projects feature allows your project to run in advanced cloud environments such as Kubernetes and Databricks. Scaling your ML job seamlessly is one of the main selling points of a platform such as MLflow.

As you have seen from the current section, the **MLflow project** module allows the execution of a reproducible ML job that is treated as a self-contained project.

Exploring MLflow tracking

The **MLflow tracking** component is responsible for observability. The main features of this module are the logging of metrics, artifacts, and parameters of an MLflow execution. It provides vizualisations and artifact management features.

In a production setting, it is used as a centralized tracking server implemented in Python that can be shared by a group of ML practitioners in an organization. This enables improvements in ML models to be shared within the organization.

In *Figure 1.4*, you can see an interface that logs all the runs of your model and allows you to log your experiment's observables (metrics, files, models and artifacts). For each run, you can look and compare the different metrics and parameters of your module.

It addresses common pain points when model developers are comparing different iterations of their models on different parameters and settings.

The following screenshot presents the different metrics for our last run of the previous model:

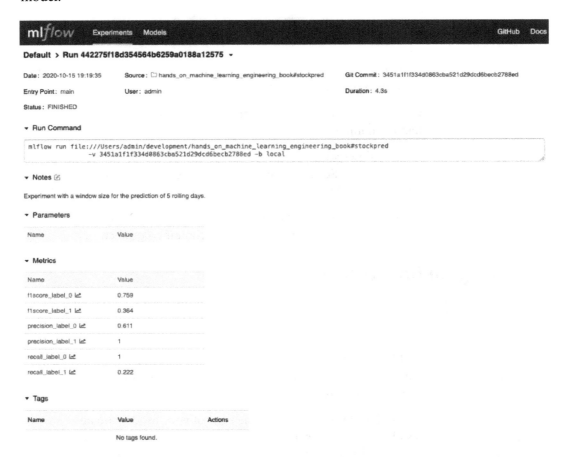

Figure 1.4 – Sample of the MLFlow interface/UI

MLflow allows the inspection of arbitrary artifacts associated with each model and its associated metadata, allowing metrics of different runs to be compared. You can see the RUN IDs and the Git hash of the code that generated the specific run of your experiment:

Figure 1.5 – Inspecting logged model artifacts

In your current directory of `stockpred`, you can run the following command to have access to the results of your runs:

```
mlflow ui
```

Running the MLflow UI locally will make it available at the following URL: `http://127.0.0.1:5000/`.

In the particular case of the runs shown in the following screenshot, we have a named experiment where the parameter of the size of the window in the previous example was tweaked. Clear differences can be seen between the performance of the algorithms in terms of F1 score:

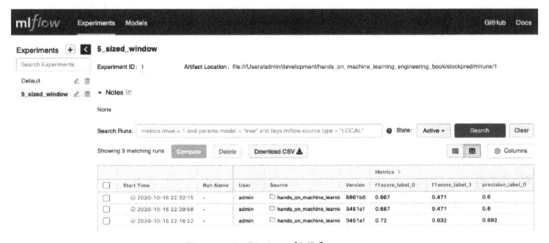

Figure 1.6 – Listing of MLflow runs

Another very useful feature of MLFlow tracking is the ability to compare between different runs of jobs:

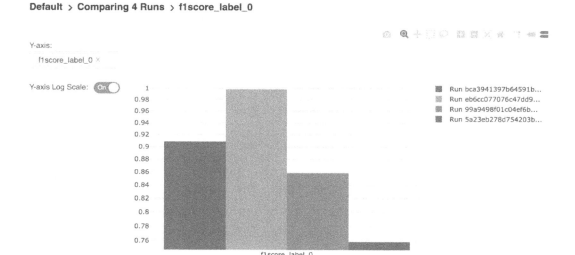

Figure 1.7 – Comparison of F1 metrics of job runs

This preceding visualization allows a practitioner to make a decision as to which model to use in production or whether to iterate further.

Exploring MLflow Models

MLflow Models is the core component that handles the different model flavors that are supported in MLflow and intermediates the deployment into different execution environments.

We will now delve into the different models supported in the latest version of MLflow.

As shown in the *Getting started with MLflow* section, MLflow models have a specific serialization approach for when the model is persisted in its internal format. For example, the serialized folder of the model implemented on the stockpred project would look like the following:

```
├── MLmodel
├── conda.yaml
└── model.pkl
```

Internally, MLflow sklearn models are persisted with the conda files with their dependencies at the moment of being run and a pickled model as logged by the source code:

```
artifact_path: model_random_forest
flavors:
  python_function:
    env: conda.yaml
    loader_module: mlflow.sklearn
    model_path: model.pkl
    python_version: 3.7.6
  sklearn:
    pickled_model: model.pkl
    serialization_format: cloudpickle
    sklearn_version: 0.23.2
run_id: 22c91480dc2641b88131c50209073113
utc_time_created: '2020-10-15 20:16:26.619071'
~
```

MLflow, by default, supports serving models in two flavors, namely, as a python_ function or in sklearn format. The flavors are basically a format to be used by tools or environments serving models.

A good example of using the preceding is being able to serve your model without any extra code by executing the following command:

```
mlflow models serve -m ./mlruns/0/
b9ee36e80a934cef9cac3a0513db515c/artifacts/model_random_forest/
```

You have access to a very simple web server that can run your model. Your model prediction interface can be executed by running the following command:

```
curl http://127.0.0.1:5000/invocations -H 'Content-Type:
application/json' -d '{"data":[[1,1,1,1,0,1,1,1,0,1,1,1,0,0]]
}' [1]%
```

The response to the API call to our model was 1; as defined in our predicted variable, this means that in the next reading, the stock will move up.

The final few steps outline how powerful MLflow is as an end-to-end tool for model development, including for the prototyping of REST-based APIs for ML services.

The MLflow Models component allows the creation of custom-made Python modules that will have the same benefits as the built-in models, as long as a prediction interface is followed.

Some of the notable model types supported will be explored in upcoming chapters, including the following:

- XGBoost model format
- R functions
- H2O model
- Keras
- PyTorch
- Sklearn
- Spark MLib
- TensorFlow
- Fastai

Support for the most prevalent ML types of models, combined with its built-in capability for on-premises and cloud deployment, is one of the strongest features of MLflow Models. We will explore this in more detail in the deployment-related chapters.

Exploring MLflow Model Registry

The model registry component in MLflow gives the ML developer an abstraction for model life cycle management. It is a centralized store for an organization or function that allows models in the organization to be shared, created, and archived collaboratively.

The management of the model can be made with the different APIs of MLflow and with the UI. *Figure 1.7* demonstrates the Artifacts UI in the tracking server that can be used to register a model:

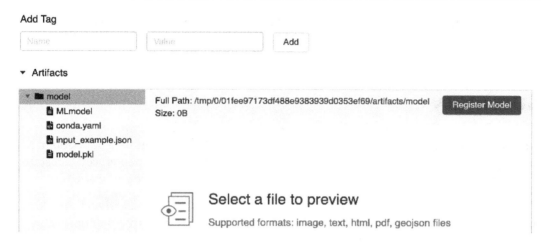

Figure 1.8 – Registering a model as an artifact

Upon registering the model, you can annotate the registered model with the relevant metadata and manage its life cycle. One example is to have models in a staging pre-production environment and manage the life cycle by sending the model to production:

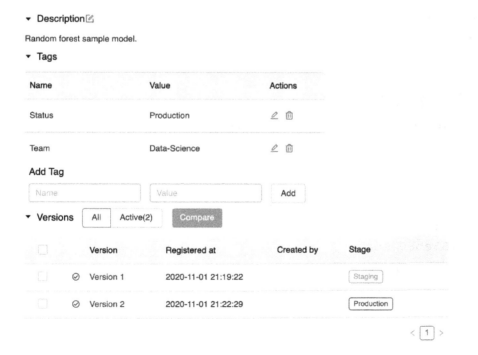

Figure 1.9 – Managing different model versions and stages

The model registry module will be explored further in the book, with details on how to set up a centralized server and manage ML model life cycles, from conception through to phasing out a model.

Summary

In this chapter, we introduced MLflow, and explored some of the motivation behind adopting a ML platform to reduce the time from model development to production in ML development. With the knowledge and experience acquired in this chapter, you can start improving and making your ML development workflow reproducible and trackable.

We delved into each of the important modules of the platform: projects, models, trackers, and model registry. A particular emphasis was given to practical examples to illustrate each of the core capabilities, allowing you to have a hands-on approach to the platform. MLflow offers multiple out-of-the-box features that will reduce friction in the ML development life cycle with minimum code and configuration. Out-of-the-box metrics management, model management, and reproducibility are provided by MLflow.

We will build on this introductory knowledge and expand our skills and knowledge in terms of building practical ML platforms in the rest of the chapters.

We briefly introduced in this chapter the use case of stock market prediction, which will be used in the rest of the book. In the next chapter, we will focus on defining rigorously the ML problem of stock market prediction.

Further reading

In order to enhance your knowledge, you can consult the documentation available at the following links:

- Reference information for MLflow is available here: `https://www.mlflow.org/docs/latest/`

- Review notes on ML platforms: `https://medium.com/nlauchande/review-notes-of-ml-platforms-uber-michelangelo-e133eb6031da`

- MLflow technical paper: `https://www-cs.stanford.edu/people/matei/papers/2018/ieee_mlflow.pdf`

2
Your Machine Learning Project

The approach of this book is to iterate through a practical business project – namely, stock market prediction – and, with this use case, explore through the different chapters the different features of MLflow. We will use a structured approach to frame a machine learning problem and project. A sample pipeline will be created and used to iterate and evolve the project in the remainder of the book.

Using a structured framework to describe a machine learning problem helps the practitioner to reason more efficiently about the different requirements of the machine learning pipeline. We will present a practical pipeline using the requirements elicited during framing.

Specifically, we will cover the following sections in this chapter:

- Exploring the machine learning process
- Framing the machine learning problem
- Introducing the stock market prediction problem
- Developing your machine learning baseline pipeline

Technical requirements

For this chapter, you will need the following prerequisites:

- The latest version of Docker installed on your machine. If you don't already have it installed, please follow the instructions at `https://docs.docker.com/get-docker/`.

- Access to a Bash terminal (Linux or Windows).

- Access to a browser.

- Python 3.5+ installed.

- MLflow installed locally as described in *Chapter 1, Introducing MLflow*.

Exploring the machine learning process

In this chapter, we will begin by describing the problem that we will solve throughout the book. We aim to focus on machine learning in the context of stock trading.

Machine learning can be defined as the process of training a software artifact – in this case, a model to make relevant predictions in a problem. Predictions are used to drive business decisions, for instance, which stock should be bought or sold or whether a picture contains a cat or not.

Having a standard approach to a machine learning project is critical for a successful project. The typical iteration of a machine learning life cycle is depicted in *Figure 2.1*:

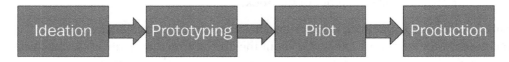

Figure 2.1 – Excerpt of the acquired data with the prediction column

Let's examine each stage in detail:

- **Ideation**: This phase involves identifying a business opportunity to use machine learning and formulating the problem.

- **Prototyping**: This involves verifying the feasibility and suitability of existing datasets to implement the planned idea.

- **Pilot**: This involves evaluating and iterating over a machine learning algorithm in order to make the decision of whether to progress or not to the subsequent phase.

- **Production deployment**: Upon successful piloting, we should be able to run the machine learning project in production and allow it to start receiving production traffic.

These high-level phases defined in *Figure 2.1* are definitely not static and are generally iterative, with dynamic movement between the phases to refine and improve.

It should be noted that machine learning is not the solution to every problem. Before deciding on machine learning, a deep evaluation needs to be made of the problem or project at hand to decide whether or not to apply machine learning.

There are simple scenarios where a machine learning solution is a good candidate, for instance, in the following cases:

- When simple rules and heuristics are not enough to solve the problem at hand.

- It's very expensive to solve the problem with the current state-of-the-art solutions.

- When the rules and code to solve a problem are very intricate and hard to maintain.

The advent of machine learning brought changes to various areas; one of these areas is the **finance** field. Before computers, finance was based on trading papers in small offices with a reduced number of transactions. With computers, the situation changed; there are now millions of transactions done per day. Millions of people trade with each other without having to meet in person or have any other sort of physical contact.

In finance and the stock trading space, machine learning can be used in the following contexts:

- **Data mining**: To identify patterns in your dataset using advanced predictive analytics. It's very common for advanced data analysts to use machine learning models as part of their analysis process and to drive business decisions.

- **Pair trading**: A technique of using two pairs of stocks that are in opposite directions of the market. Basically, this works by selling or buying when each of the stocks is out of phase with its normal behavior as the markets are known to converge after some time.

- **Stock forecasting**: Simple prediction, based on the current time series of what a particular stock will be traded for at a point in the future.

- **Anomaly detection**: Detecting abnormal situations in the market reveals itself to be very important to prevent auto-trading systems from operating on days when the market is anomalous.

- **Sentiment analysis**: The stock market is known to be driven mostly by sentiments of participants around companies and businesses. This technique uses, for instance, messages on social media or other mediums, using natural language processing techniques to gauge sentiment (for example, positive, negative, or neutral).

Next, we'll look at framing the problem for machine learning.

Framing the machine learning problem

Machine learning problem framing, as defined in this section, is a technique and methodology to help specify and contextualize a machine learning problem in such a way that an engineering solution can be implemented. Without a solid approach to tackling machine learning problems, it can become very hard to extract the real value of the undertaking.

We will draw inspiration from the approaches of companies such as Amazon and Google, which have been successfully applying the technique of machine learning problem framing.

The machine learning development process is highly based on the scientific method. We undergo different stages of stating a goal, data collection, hypothesis testing, and conclusion. It's expected that we will cycle through the different stages of the workflow until either a good model is identified or it becomes apparent that it's impossible to develop one.

The following subsections depict the framework that we will use in the rest of the book to elicit a machine learning problem-solving framework.

Problem statement

Understanding the problem that we are solving is very important before we attempt anything else. Defining a problem statement for the problem at hand is a clear way to define your problem. The following are reasonable examples of a machine learning problem statement:

- Predict whether the stock market for a specific ticker will go up tomorrow.
- Predict whether we have a cat in a picture.

In this part of the process, we specify the specific type of machine learning problem that we are aiming to solve. The following are common types of problems:

- **Classification**: This type of problem requires you to predict a label or class as the output of your model, for example, classifying whether an email text is spam or not spam. It can be a binary classification or a multiclass classification. A good example of a multiclass variant is classifying the digits given the handwriting.

- **Regression**: This is if you need, for instance, to have a model predict a numeric value from a training dataset. Good examples are predicting temperature based on atmospheric features and predicting the exact value in dollars of a given stock.

- **Clustering**: Consists of discovering natural groupings of data when you don't have labels to train models. It groups similar data points into the same groups using a distance metric. Clustering can be used to identify fraudulent transactions as they won't belong to an existing grouping.

- **Generative model**: This is a novel type of machine learning where new data is generated based on the existing data and is statistically similar to the input data. It's widely used in modern language models such as GPT-3 from OpenAI.

- **Reinforcement learning**: A type of machine learning where the agent/model interacts with the environment to learn optimal behavior to maximize the reward. One famous application is the AlphaGo agent from Google DeepMind, which was able to outperform the best human player at the board game Go. A very important application is to train an automated trading stock agent using profitability as the reward.

> **Important note**
> There are multiple taxonomies for machine learning problems; the listing in this section is by no means exhaustive. This list is relevant to the examples in the book.

Success and failure definition

Getting the success definition from a business is very important to put your problem in perspective. For instance, for the stock market case, the following can be outlined as a desirable outcome: "Using the model of this project, we would like to *improve* the *profitability* up to 56% of our daily *trading* operations as they are currently profitable in 50% of the daily trades."

So, if we have successful trades only 30% of the time, the model is clearly a failure. It will generally be dependent on business criteria.

Having a business metric as a success definition, instead of a technical metric such as accuracy, precision, or recall, helps keep the solution aligned to tangible goals to an organization.

Model output

The model output in the context of machine learning problem framing is dependent on the type of problem that you are solving. For instance, a regression model output will be a specific number (for example, 10.5 as a stock prediction) and a classification will return a label (for example, `true` when a cat is detected) and a probabilistic threshold.

Your model output should definitely be related to your outcome definition.

Output usage

Stating how your prediction will be used helps unveil the reasoning for the model development, helping the developers to get context and ownership of the problem, for instance, deciding whether you will use your prediction directly in a UI or use it to feed an upstream system.

Heuristics

Problems solved with machine learning can be approximated by rules and heuristics. Having a heuristic as a starting point for a machine learning pipeline is generally a recommended approach to starting a project.

For instance, a valid heuristic for the stock trading prediction problem would be using the last day's prediction as the first baseline production heuristic. The goal of the model developer is then to beat the baseline with a better model.

Data layer definition

Defining precisely the input and output in the context of your model helps clarify and guide the development of your machine learning problem.

Data sources

This section on framing the problem comprises identifying raw data sources and documenting them during the problem framing process. Examples of raw data sources are public or proprietary datasets and their definitions and data dictionaries.

It's important to identify whether the data is labeled or not at this stage, and the effort that will be needed to label the data.

Model input/output

Model development involves defining the precise inputs to be used in the model. Given all the data sources available, specifying the model input or feature sets becomes essential in order to execute your pipeline. Alongside your input, the desired target should be defined.

The model input/output is better framed as a table, as in *Figure 2.2*, in order to facilitate reasoning and model implementation. One row of example data is added for clarity:

Sepal Length – Input	Sepal Length - Input	Sepal Length - Input	Sepal Length – Input	Class(Label)
Integer	Integer	Integer	Integer	String(Label)
10	20	30	40	Iris Setosa

Figure 2.2 – Example of documenting the inputs/outputs of a model

Next, we'll look at using the machine learning problem framing technique on the stock trading scenario that we will work on throughout the book.

Introducing the stock market prediction problem

The scenario that we will cover in the remaining chapters of the book is of the hypothetical company **PsyStock LLC**, which provides a platform for amateur traders, providing APIs and UIs to solve different predictions in the context of stock prediction.

As machine learning practitioners and developers, we should be able to build a platform that will allow a team of data scientists to quickly develop, test, and bring into production machine learning projects.

We will apply and frame the problems initially so we can build our platform upon the basis of the definitions of the problems. It should be noted that the problem framing will evolve as we learn more about the problem: the initial framing will give us guidance on the problem spaces that we will be tackling.

The following are the core projects that we will use as references in the rest of the book for machine learning development in MLflow.

Stock movement predictor

This is the project for the first API that the company PsyStock LLC will provide to its customers. This API will return `true` if a given stock ticker will go up in the stock market and `false` if not.

Problem statement

The problem statement is to *predict through a machine learning classifier whether the market will go up or not in a single day.*

Success and failure definition

Success, in this case, is defined by *the percentage of days in a month where the system predicted the correct direction of the market.* Success is basically whether the system is accurate, from a market direction perspective, more than 60% of the time – basically, the expected value of being better than the random baseline.

Model output

The model output is 1 for an increase in value of a stock ticker and 0 for non-increase.

Output usage

The output of the model will be used to provide the Rest API with a `true` or `false` value based on a defined threshold of accuracy on the classification.

The expected latency for this problem is *under 1,000 milliseconds.*

Heuristics

The simplest heuristic to solve this problem is using a random predictor for each input, with an equal probability of the market going up or down.

Data layer definition

Let's define each part of our data layer.

Data sources

Historical stock market datasets as provided by the Yahoo Finance public API.

Model input/output

Let's look at the input and output, next:

- **Input**: Historical end-of-the-day price of a given ticker for the last 10 days.

- **Output**: 1 for increasing and 0 for not increasing in the next period.

The following table shows the input/output data of a model:

Day N-N Input	Day N-9 Input	...	Day N-1 Input	Class(Label)
Float	Float	Float	Float	Integer
103.2	203.1	...	200.1	1

Figure 2.3 – Example of documenting the input/outputs of a model

Sentiment analysis of market influencers

The sentiment machine learning pipeline will *predict whether the sentiment over a stock ticker is positive or negative on social media* and provide it as an API to the users of the machine learning platform that we are developing in this book.

Problem statement

To predict whether a given stock ticker has positive sentiment for the current day of relevant market influencers on Twitter selected by PsyStock LLC.

Success and failure definition

Success, in this case, is a bit harder to define, as the fact of a sentiment being positive can't exactly be tracked to a market metric. The definition of success on this particular prediction problem should be a proxy for how many times a user is a repeat user of the API.

Model output

The model output is basically a number matching the polarity of the tweet – positive, negative, or neutral sentiment – of a ticker.

Output usage

The output of this system will be used in a Rest API that will be provided on request, with the number of positive, neutral, and negative polarities for a given ticker.

Heuristics

A very simple heuristic to implement as a baseline for this problem is to count the number of times that the words up and down are used. Whichever word is more frequent for a ticker, its value will be used for polarity. If the percentage of frequency between the two words is less than 10%, we will assume that the ticker sentiment was neutral.

Data layer definition

The most readily available raw data for this problem is social media. Twitter provides an easy-to-consume and search API where we can search by ticker and influencer handle.

Data sources

The source data will be acquired through the Twitter API to search for tickers given an updatable list of market influencers.

Model input/output

The system that will serve the model will receive a tweet and return the classified sentiment (positive, negative, or neutral):

Tweet Input	Class(Label)
String	Integer
"It seems that $AMZN will break up positively "	Positive

Figure 2.4 – Example of documenting the input/outputs of a model

In this section, we just defined the initial problems that we will tackle, end to end, throughout the book. We will be refining the definition of the problem frame and improving it when necessary and updating the problem frame as required.

In the next section, we'll look at using the definition of the heuristics during problem framing to create the first base pipeline that we will work on improving.

Developing your machine learning baseline pipeline

For our machine learning platform, we will start with a very simple, heuristic-based pipeline, in order to get the infrastructure of your end-to-end system working correctly and an environment where the machine learning models can iterate on it.

> **Important note**
> It is critical that the technical requirements are correctly installed in your local machine to follow along. The assumption on this section is that you have MLflow and Docker installed as per the *Technical requirements* section.

By the end of this section, you will be able to create our baseline pipeline. The baseline pipeline value is to enable rapid iteration to the model developers. So, basically, an end-to-end infrastructure with placeholders for training and model serving will be made available to the development team. Since it's all implemented in MLflow, it becomes easy to have specialization and focus of the different types of teams involved in a machine learning project. The engineering teams will focus on improving the pipeline while the data science-oriented teams will have a way to rapidly test and evaluate their models:

1. Implement the heuristic model in MLflow.

   In the following block of code, we create the RandomPredictor class, a bespoke predictor that descends from the mlflow.pyfunc.PythonModel class. The main predict method returns a random number between 0 and 1:

   ```
   import mlflow
   class RandomPredictor(mlflow.pyfunc.PythonModel):
     def __init__(self):
       pass

     def predict(self, context, model_input):
   ```

```
      return model_input.apply(lambda column: random.
randint(0,1))
```

We use a specific functionality of creating a custom model in MLflow; more details about custom models can be found at `https://mlflow.org/docs/latest/python_api/mlflow.pyfunc.html#pyfunc-create-custom`.

2. Save the model in MLflow.

    The following block of code saves the model with the name `random_model` in a way that can be retrieved later on. It registers within the MLflow registry in the local filesystem:

    ```
    model_path = "random_model"
    baseline_model = RandomPredictor()
    mlflow.pyfunc.save_model(path=model_path, python_
    model=random_model)
    ```

    At this stage, we basically instantiate the model and store it on the model registry as configured by the local environment.

3. Run your `mlflow` job:

    ```
    mlflow run .
    ```

4. Start the serving API:

    ```
    mlflow models serve -m ./mlruns/0/
    b9ee36e80a934cef9cac3a0513db515c/artifacts/random_model/
    ```

5. Test the API of your model.

    You have access to a very simple Flask server that can run your model. You can test the execution by running a `curl` command in your server:

    ```
    curl http://127.0.0.1:5000/invocations -H 'Content-Type:
    application/json' -d '{"data":[[1,1,1,1,0,1,1,1,0,1,1,1,0
    ,0]]}' [1]%
    ```

At this stage, we have a baseline dummy model that the goal of the model developers team is now to improve upon. This same pipeline will be used in the next chapter to build a data science development environment with the initial algorithms of the platform being built in this book.

Summary

In this chapter, we introduced the machine learning problem framing approach, and explored some of the motivation behind adopting this framework.

We introduced the stock market prediction machine learning platform and our initial set of prediction problems using the ML problem framing methodology.

We briefly introduced in this chapter the use case of a stock market prediction basic pipeline that will be used in the rest of the book.

In the next chapter, we will focus on creating a data science development environment with MLflow using the definitions of the problem made in this chapter.

Further reading

In order to further your knowledge, you can consult the documentation at the following links:

- Reference information on the Google machine learning problem framing framework: `https://developers.google.com/machine-learning/problem-framing`

- Reference information on the Amazon machine learning framing framework: `https://docs.aws.amazon.com/wellarchitected/latest/machine-learning-lens/ml-problem-framing.html`

Section 2: Model Development and Experimentation

This section covers the process of end-to-end model development with MLflow, including experimentation management through MLflow, targeting the algorithm development phase in machine learning. You will learn how to use MLflow to iteratively develop your machine learning model and manage it with MLflow.

The following chapters are covered in this section:

- *Chapter 3, Your Data Science Workbench*
- *Chapter 4, Experiment Management in MLflow*
- *Chapter 5, Managing Models with MLflow*

3

Your Data Science Workbench

In this chapter, you will learn about MLflow in the context of creating a local environment so that you can develop your machine learning project locally with the different features provided by MLflow. This chapter is focused on machine learning engineering, and one of the most important roles of a machine learning engineer is to build up an environment where model developers and practitioners can be efficient. We will also demonstrate a hands-on example of how we can use workbenches to accomplish specific tasks.

Specifically, we will look at the following topics in this chapter:

- Understanding the value of a data science workbench
- Creating your own data science workbench
- Using the workbench for stock prediction

Technical requirements

For this chapter, you will need the following prerequisites:

- The latest version of Docker installed on your machine. If you don't already have it installed, please follow the instructions at `https://docs.docker.com/get-docker/`.

  The latest version of Docker Compose installed. If you don't already have it installed, please follow the instructions at `https://docs.docker.com/compose/install/`.

- Access to Git in the command line, and installed as described in this **Uniform Resource Locator** (**URL**): `https://git-scm.com/book/en/v2/Getting-Started-Installing-Git`.

- Access to a `bash` terminal (Linux or Windows).

- Access to a browser.

- Python 3.5+ installed.

- MLflow installed locally, as described in *Chapter 1, Introducing MLflow*.

Understanding the value of a data science workbench

A data science workbench is an environment to standardize the machine learning tools and practices of an organization, allowing for rapid onboarding and development of models and analytics. One critical machine learning engineering function is to support data science practitioners with tools that empower and accelerate their day-to-day activities.

In a data science team, the ability to rapidly test multiple approaches and techniques is paramount. Every day, new libraries and open source tools are created. It is common for a project to need more than a dozen libraries in order to test a new type of model. These multitudes of libraries, if not collated correctly, might cause bugs or incompatibilities in the model.

Data is at the center of a data science workflow. Having clean datasets available for developing and evaluating models is critical. With an abundance of huge datasets, specialized big data tooling is necessary to process the data. Data can appear in multiple formats and velocities for analysis or experimentation, and can be available in multiple formats and mediums. It can be available through files, the cloud, or **REpresentational State Transfer** (**REST**) **application programming interfaces** (**APIs**).

Data science is mostly a collaborative craft; it's part of a workflow to share models and processes among team members. Invariably, one pain point that emerges from that activity is the cross-reproducibility of model development jobs among practitioners. Data scientist A shares a training script of a model that assumes version 2.6 of a library, but data scientist B is using version 2.8 environment. Tracing and fixing the issue can take hours in some cases. If this problem occurs in a production environment, it can become extremely costly to the company.

When iterating—for instance—over a model, each run contains multiple parameters that can be tweaked to improve it. Maintaining traceability of which parameter yielded a specific performance metric—such as accuracy, for instance—can be problematic if we don't store details of the experiment in a structured manner. Going back to a specific batch of settings that produced a better model may be impossible if we only keep the latest settings during the model development phase.

The need to iterate quickly can cause many frustrations when translating prototype code to a production environment, where it can be executed in a reliable manner. For instance, if you are developing a new trading model in a Windows machine with easy access to **graphics processing units (GPUs)** for inference, your engineering team member may decide to reuse the existing Linux infrastructure without GPU access. This leads to a situation where your production algorithm ends up taking 5 hours and locally runs in 30 seconds, impacting the final outcome of the project.

It is clear that a data science department risks systemic technical pain if issues related to the environment and tools are not addressed upfront. To summarize, we can list the following main points as described in this section:

- Reproducibility friction
- The complexity of handling large and varied datasets
- Poor management of experiment settings
- Drift between local and production environments

A data science workbench addresses the pain points described in this section by creating a structured environment where a machine learning practitioner can be empowered to develop and deploy their models reliably, with reduced friction. A no-friction environment will allow highly costly model development hours to be focused on developing and iterating models, rather than on solving tooling and data technical issues.

After having delved into the motivation for building a data science workbench for a machine learning team, we will next start designing the data science workbench based on known pain points.

Creating your own data science workbench

In order to address common frictions for developing models in data science, as described in the previous section, we need to provide data scientists and practitioners with a standardized environment in which they can develop and manage their work. A data science workbench should allow you to quick-start a project, and the availability of an environment with a set of starting tools and frameworks allows data scientists to rapidly jump-start a project.

The data scientist and machine learning practitioner are at the center of the workbench: they should have a reliable platform that allows them to develop and add value to the organization, with their models at their fingertips.

The following diagram depicts the core features of a data science workbench:

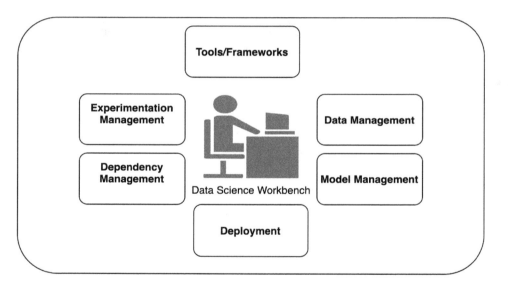

Figure 3.1 – Core features of a data science workbench

In order to think about the design of our data science workbench and based on the diagram in *Figure 3.1*, we need the following core features in our data science workbench:

- **Dependency Management**: Having dependency management built into your local environment helps in handling reproducibility issues and preventing library conflicts between different environments. This is generally achieved by using environment managers such as Docker or having environment management frameworks available in your programming language. MLflow provides this through the support of Docker- or Conda-based environments.

- **Data Management**: Managing data in a local environment can be complex and daunting if you have to handle huge datasets. Having a standardized definition of how you handle data in your local projects allows others to freely collaborate on your projects and understand the structures available.

- **Model Management**: Having the different models organized and properly stored provides an easy structure to be able to work through many ideas at the same time and persist the ones that have potential. MLflow helps support this through the model format abstraction and **Model Registry** component to manage models.

- **Deployment**: Having a development environment aligned with the production environment where the model will be serviced requires deliberation in the local environment. The production environment needs to be ready to receive a model from a model developer, with the least possible friction. This smooth deployment workflow is only possible if the local environment is engineered correctly.

- **Experimentation Management**: Tweaking parameters is the most common thing that a machine learning practitioner does. Being able to keep abreast of the different versions and specific parameters can quickly become cumbersome for the model developer.

> **Important note**
> In this section, we will implement the foundations of a data science workbench from scratch with MLflow, with support primarily for local development. There are a couple of very opinionated and feature-rich options provided by cloud providers such as **Amazon Web Services** (**AWS**) Sagemaker, Google AI, and **Azure Machine Learning** (**Azure ML**).

Machine learning engineering teams have freedom in terms of the use cases and technologies that the team they are serving will use.

The following steps demonstrate a good workflow for development with a data science workbench:

- The model developer installs the company workbench package through an installer or by cloning the repository.

- The model developer runs a command to start a project.

- The model developer chooses a set of options based on configuration or a prompt.

- The basic scaffolding is produced with specific folders for the following items:

  a) `Data`: This will contain all the data assets of your current project

  b) `Notebooks`: To hold all the iterative development notebooks with all the steps required to produce the model

  c) `Model`: A folder that contains the binary model or a reference to models, potentially in binary format

  d) `Source Code`: A folder to store the structured code component of the code and reusable libraries

  e) `Output`: A folder for any specific outputs of the project—for instance, visualizations, reports, or predictions

- A project folder is created with the standards for the organization around packages, dependency management, and tools.

- The model developer is free to iterate and create models using supported tooling at an organizational level.

Establishing a data science workbench provides a tool for acceleration and democratization of machine learning in the organization, due to standardization and efficient adoption of machine learning best practices.

We will start our workbench implementation in our chapter with sensible components used industrywide.

Building our workbench

We will have the following components in the architecture of our development environment:

- **Docker/Docker Compose**: Docker will be used to handle each of the main component dependencies of the architecture, and Docker Compose will be used as a coordinator between different containers of software pieces. The advantage of having each component of the workbench architecture in Docker is that neither element's libraries will conflict with the other.

- **JupyterLab**: The de facto environment to develop data science code and analytics in the context of machine learning.

- **MLflow**: MLflow is at the cornerstone of the workbench, providing facilities for experiment tracking, model management, registry, and deployment interface.

- **PostgreSQL database**: The PostgreSQL database is part of the architecture at this stage, as the storage layer for MLflow for backend metadata. Other relational databases could be used as the MLflow backend for metadata, but we will use PostgreSQL.

Our data science workbench design can be seen in the following diagram:

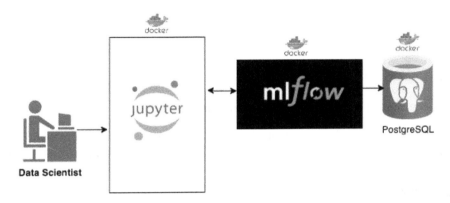

Figure 3.2 – Our data science workbench design

Figure 3.2 illustrates the layout of the proposed components that will underpin our data science workbench.

The usual workflow of the practitioner, once the environment is up and running, is to develop their code in Jupyter and run their experiments with MLflow support. The environment will automatically route to the right MLflow installation configured to the correct backend, as shown in *Figure 3.2*.

> **Important note**
> Our data science workbench, as defined in this chapter, is a complete local environment. As the book progresses, we will introduce cloud-based environments and link our workbench to shared resources.

A sample layout of the project is available in the following GitHub folder:

https://github.com/PacktPublishing/Machine-Learning-Engineering-with-MLflow/tree/master/Chapter03/gradflow

You can see a representation of the general layout of the workbench in terms of files here:

```
├── Makefile
├── README.md
```

```
├── data
├── docker
├── docker-compose.yml
├── docs
├── notebooks
├── requirements.txt
├── setup.py
├── src
├── tests
└── tox.ini
```

The main elements of this folder structure are outlined here:

- `Makefile`: This allows control of your workbench. By issuing commands, you can ask your workbench to set up a new environment notebook to start MLflow in different formats.

- `README.md`: A file that contains a sample description of your project and how to run it.

- `data` folder: A folder where we store the datasets used during development and mount the data directories of the database when running locally.

- `docker`: A folder that encloses the Docker images of the different subsystems that our environment consists of.

- `docker-compose.yml`: A file that contains the orchestration of different services in our workbench environment—namely: Jupyter Notebooks, MLflow, and PostgreSQL to back MLflow.

- `docs`: Contains relevant project documentation that we want persisted for the project.

- `notebooks`: A folder that contains the notebook information.

- `requirements.txt`: A requirements file to add libraries to the project.

- `src`: A folder that encloses the source code of the project, to be updated in further phases of the project.

- `tests`: A folder that contains end-to-end testing for the code of the project.

- `tox.ini`: A templated file that controls the execution of unit tests.

We will now move on to using our own development environment for a stock-prediction problem, based on the framework we have just built.

Using the workbench for stock prediction

In this section, we will use the workbench step by step to set up a new project. Follow the instructions step by step to start up your environment and use the workbench for the stock-prediction project.

> **Important note**
> It is critical that all packages/libraries listed in the *Technical requirements* section are correctly installed on your local machine to enable you to follow along.

Starting up your environment

We will move on next to exploring your own development environment, based on the development environment shown in this section. Please execute the following steps:

1. Copy the contents of the project available in `https://github.com/ PacktPublishing/Machine-Learning-Engineering-with-MLflow/ tree/master/Chapter03/gradflow`.

2. Start your local environment by running the following command:

    ```
    make
    ```

3. Inspect the created environments, like this:

    ```
    $ docker ps
    ```

    The following screenshot presents three Docker images: the first for Jupyter, the second for MLflow, and the third for the PostgreSQL database. The status should show `Up x minutes`:

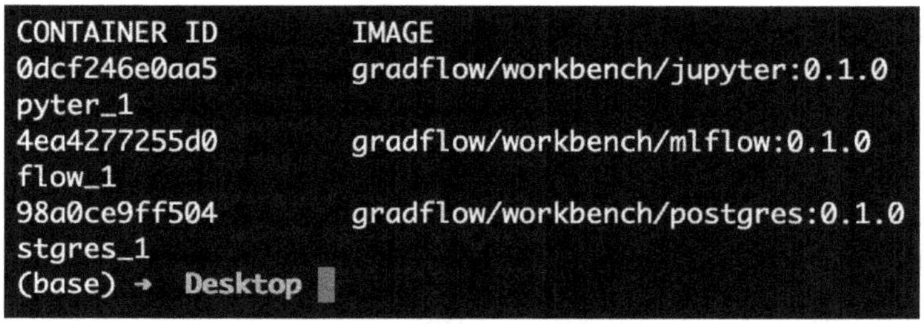

```
CONTAINER ID          IMAGE
0dcf246e0aa5          gradflow/workbench/jupyter:0.1.0
pyter_1
4ea4277255d0          gradflow/workbench/mlflow:0.1.0
flow_1
98a0ce9ff504          gradflow/workbench/postgres:0.1.0
stgres_1
(base) → Desktop
```

Figure 3.3 – Running Docker images

The usual ports used by your workbench are listed as follows: Jupyter serves in port 8888, MLflow serves in port 5000, and PostgreSQL serves in port 5432.

In case any of the containers fail, you might want to check if the ports are used by different services. If this is the case, you will need to turn off all of the other services.

Check your Jupyter Notebooks environment at http://localhost:8888, as illustrated in the following screenshot:

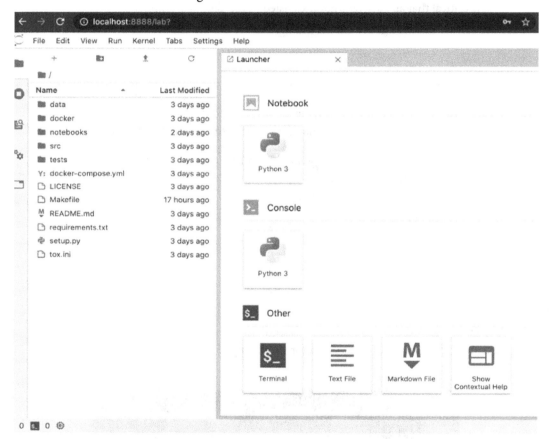

Figure 3.4 – Running Jupyter environment

You should have a usable environment, allowing you to create new notebooks file in the specified folder.

Check your MLflow environment at http://localhost:5000, as illustrated in the following screenshot:

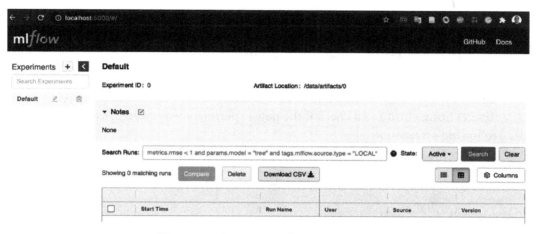

Figure 3.5 – Running MLflow environment

Figure 3.5 shows your experiment tracker environment in MLflow that you will use to visualize your experiments running in MLflow.

Run a sample experiment in MLflow by running the `notebook` file available in `/notebooks/mlflow_sample.ipynb`, as illustrated in the following screenshot:

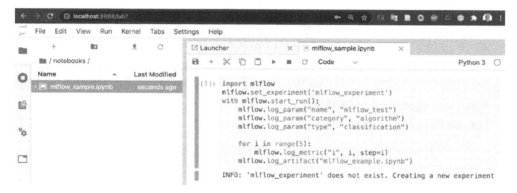

Figure 3.6 – Excerpt of mlflow_sample code

The code in *Figure 3.6* imports MLflow and creates a dummy experiment manually, on the second line, using `mlflow.set_experiment('mlflow_experiment')`.

The `with mlflow.start_run()` line is responsible for starting and tearing down the experiment in MLflow.

In the three following lines, we log a couple of string-type test parameters, using the `mlflow.log_param` function. To log numeric values, we will use the `mlflow.log_metric` function.

Finally, we also log the entire file that executed the function to ensure traceability of the model and code that originated it, using the `mlflow.log_artifact("mlflow_example.ipynb")` function.

Check the sample runs, to confirm that the environment is working correctly. You should go back to the MLflow **user interface** (**UI**) available at `http://localhost:5000` and check if the new experiment was created, as shown in the following screenshot:

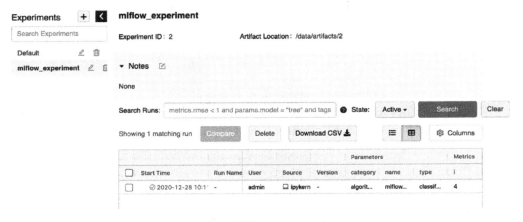

Figure 3.7 – MLflow test experiment

Figure 3.7 displays the additional parameters that we used on our specific experiment and the specific metric named `i` that is visible in the **Metrics** column.

Next, you should click on the experiment created to have access to the details of the run we have executed so far. This is illustrated in the following screenshot:

Figure 3.8 – MLflow experiment details

Apart from details of the metrics, you also have access to the `mlflow_example` notebook file at a specific point in time.

At this stage, you have your environment running and working as expected. Next, we will update it with our own algorithm; we'll use the one we created in *Chapter 2, Your Machine Learning Project*.

Updating with your own algorithms

Let's update the notebook file that we created in *Chapter 2, ML Problem Framing*, and add it to the notebook folder on your local workbench. The code excerpt is presented here:

```
import mlflow
class RandomPredictor(mlflow.pyfunc.PythonModel):
  def __init__(self):
    pass


  def predict(self, context, model_input):
    return model_input.apply(lambda column: random.
randint(0,1))
```

Under the `notebook` folder in the `notebooks/stockpred_randomizer.ipynb` file, you can follow along with the integration of the preceding code excerpt in our recently created data science workbench. We will proceed as follows:

1. We will first import all the dependencies needed and run the first cell of the notebook, as follows:

```
import random
import mlflow
from mlflow.pyfunc.model import PythonModel
```

Figure 3.9 – MLflow experiment details

2. Let's declare and execute the class outlined in *Figure 3.9*, represented in the second cell of the notebook, as follows:

```
class RandomPredictor(PythonModel):
  def __init__(self):
    pass

  def predict(self, context, model_input):
    return model_input.apply(lambda column: random.randint(0,1))
```

Figure 3.10 – Notebook cell with the RandomPredictor class declaration

3. We can now save our model in the MLflow infrastructure so that we can test the loading of the model. `model_path` holds the folder name where the model will be saved. You need to instantiate the model in an `r` variable and use `mlflow.pyfunc.save_model` to save the model locally, as illustrated in the following code snippet:

```
# Construct and save the model
model_path = "randomizer_model"
r = RandomPredictor()
mlflow.pyfunc.save_model(path=model_path, python_model=r)
```

Figure 3.11 – Notebook demonstrating saving the model

You can see on the left pane of your notebook environment that a new folder was created alongside your files to store your models. This folder will store the Conda environment and the pickled/binarized Python function of your model, as illustrated in the following screenshot:

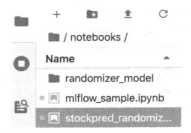

Figure 3.12 – Notebook demonstrating the saved model folder

4. Next, we can load and use the model to check that the saved model is usable, as follows:

```
# Load the model in `python_function` format
loaded_model = mlflow.pyfunc.load_model(model_path)
```

Figure 3.13 – Notebook demonstrating the saved model folder

Figure 3.14 demonstrates the creation of a random input **pandas DataFrame** and the use of `loaded_model` to predict over the input vector. We will run the experiment with the name `stockpred_experiment_days_up`, logging as a metric the number of days on which the market was up on each of the models, as follows:

```python
import pandas as pd
model_input = pd.DataFrame([range(10)])

random_predictor = RandomPredictor()

mlflow.set_experiment('stockpred_experiment_days_up')
with mlflow.start_run():
    model_output = loaded_model.predict(model_input)

    mlflow.log_metric("Days Up",model_output.sum())
    mlflow.log_artifact("stockpred_randomizer.ipynb")
```

Figure 3.14 – Notebook cell demonstrating use of the loaded model

To check the last runs of the experiment, you can look at `http://localhost:5000` and check that the new experiment was created, as illustrated in the following screenshot:

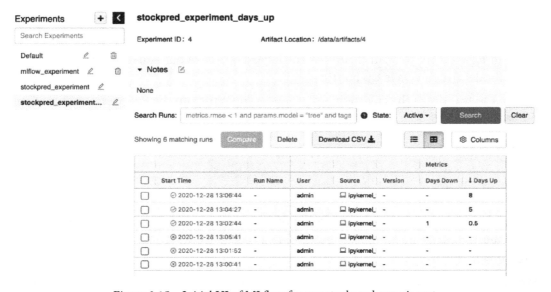

Figure 3.15 – Initial UI of MLflow for our stockpred experiment

You can now compare multiple runs of our algorithm and see differences in the **Days Up** metric, as illustrated in the following screenshot. You can choose accordingly to delve deeper on a run that you would like to have more details about:

stockpred_experiment_days_up > **Run de350b9f79f741aca8fc42b94003638c** ▾

Date : 2020-12-28 13:06:44   Source : 💻 ipykernel_launcher.py   User : admin

Duration : 108ms   Status : FINISHED

▾ Notes ☑

None

▸ Parameters

▾ Metrics

Name	Value
Days Up 📈	8

▸ Tags

▾ Artifacts

📄 stockpred_randomizer.ipynb

Figure 3.16 – Logged details of the artifacts saved

In *Figure 3.16*, you can clearly see the logged details of our run—namely, the artifact model and the **Days Up** metric.

In order to tear down the environment properly, you must run the following command in the same folder:

```
make down
```

Summary

In this chapter, we introduced the concept of a data science workbench and explored some of the motivation behind adopting this tool as a way to accelerate our machine learning engineering practice.

We designed a data science workbench, using MLflow and adjacent technologies based on our requirements. We detailed the steps to set up your development environment with MLflow and illustrated how to use it with existing code. In later sections, we explored the workbench and added to it our stock-trading algorithm developed in the last chapter.

In the next chapter, we will focus on experimentation to improve our models with MLflow, using the workbench developed in this chapter.

Further reading

In order to further your knowledge, you can consult the documentation in the following links:

- Cookiecutter documentation page: `https://cookiecutter.readthedocs.io/en/1.7.2/`

- Reference information about cookie cutters: `https://drivendata.github.io/cookiecutter-data-science/`

- The motivation behind data science workbenches: `https://dzone.com/articles/what-is-a-data-science-workbench-and-why-do-data-s#`

4

Experiment Management in MLflow

In this chapter, we will give you practical experience with stock predictions by creating different models and comparing metrics of different runs in MLflow. You will be guided in terms of how to use the MLflow experiment method so that different machine learning practitioners can share metrics and improve on the same model.

Specifically, we will look at the following topics in this chapter:

- Getting started with the experiments module
- Defining the experiment
- Adding experiments
- Comparing different models
- Tuning your model with hyperparameter optimization

At this stage, we currently have a baseline pipeline that acts based on a naïve heuristic. In this chapter, we will add to our set of skills the ability to experiment with multiple models and tune one specific model using MLflow.

We will be delving into our **Psystock** company use case of a stock trading machine learning platform introduced in *Chapter 2, Your Machine Learning Project*. In this chapter, we will add to our platform to compare multiple models and run experiments in the benchmark to be able to create a predictor for a specific stock and ticker.

In data science functions, a common methodology is to develop a model for a specific model that involves the following three steps: creating baseline models with different model types, identifying the best performant model, and predicting with the best model.

Technical requirements

For this chapter, you will need the following prerequisites:

- The latest version of Docker installed on your machine. If you don't already have it installed, please follow the instructions at `https://docs.docker.com/get-docker/`.

- The latest version of Docker Compose installed. Please follow the instructions at `https://docs.docker.com/compose/install/`.

- Access to Git in the command line and installed as described in `https://git-scm.com/book/en/v2/Getting-Started-Installing-Git`.

- Access to a bash terminal (Linux or Windows).

- Access to a browser.

- Python 3.5+ installed.

- The latest version of your machine learning installed locally and described in *Chapter 3, Your Data Science Workbench*.

Getting started with the experiments module

To get started with the technical modules, you will need to get started with the environment prepared for this chapter in the following folder: `https://github.com/PacktPublishing/Machine-Learning-Engineering-with-MLflow/tree/master/Chapter04`

You should be able, at this stage, to execute the `make` command to build up your workbench with the dependencies needed to follow along with this chapter. You need next to type the following command to move to the right directory:

```
$ cd Chapter04/gradflow/
```

To start the environment, you need to run the following command:

```
$ make
```

The entry point to start managing experimentation in **MLflow** is the experiments interface illustrated in *Figure 4.1*:

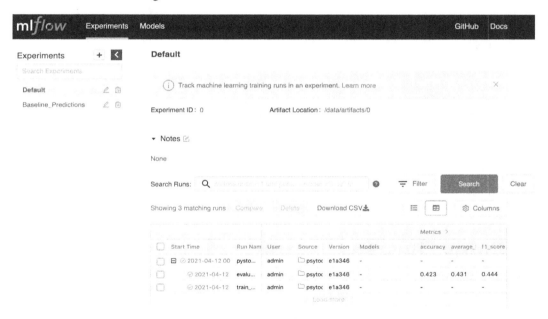

Figure 4.1 – The Experiments interface in MLflow

On the left pane (1), you can manage and create experiments, and on the right (2), you can query details of a specific experiment.

To create a new experiment, you need to click on the + button on the left pane and add the details of your experiment, as illustrated by *Figure 4.2*:

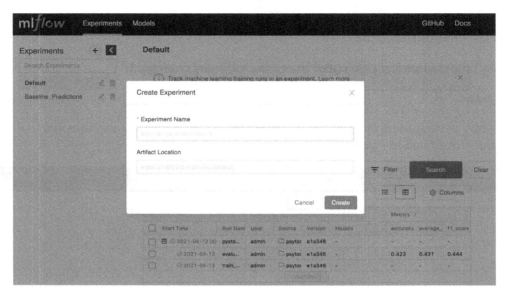

Figure 4.2 – Creating new experiments

Having introduced at a high level the tracking server and the experiment management features, we will now proceed to use the features available on our workbench to tackle the challenges of the current chapter.

Defining the experiment

Using the machine learning problem framing methodology, we will now define the main components of our stock price prediction problem as defined for the chapter:

Problem statement	Predict through a machine learning classifier whether the market for BTC(bitcoin ticker)/USD will go up or not in a single day.
Success and failure definition	Success for this particular stage of the problem will be defined by the highest performance metric of the **F Score**.

Model output	The model output is 1 for an increase in value of a stock ticker and 0 for a non-increase.
Output usage	The output of the model will be used to provide a rest API with true or false values based on a defined threshold of an F1 performance metric on the classification.
Data layer definition	Historical stock market datasets as provided by Yahoo Finance public APIs. **Input:** Historical end of the day's price of a given ticker for the last 10 days. **Output:** 1 for increasing and 0 for not increasing in the next period.

Table 4.1 – Machine learning problem framing recap

The **F-score** metric in machine learning is a measure of accuracy for binary classifiers and provides a good balance and trade-off between misclassifications (false positives or false negatives). Further details can be found on the Wikipedia page: `https://en.wikipedia.org/wiki/F-score`.

Exploring the dataset

As specified in our machine learning problem framing, we will use as input data the market observations for the period January-December 2020, as provided by the Yahoo data API.

The following code excerpt, which uses the `pandas_datareader` module available in our workbench, allows us to easily retrieve the data that we want. The complete working notebook is available at `https://github.com/PacktPublishing/Machine-Learning-Engineering-with-MLflow/blob/master/Chapter04/gradflow/notebooks/retrieve_training_data.ipynb`:

```
import pandas as pd
import numpy as np
import datetime
import pandas_datareader.data as web
```

```
from pandas import Series, DataFrame
start = datetime.datetime(2014, 1, 1)
end = datetime.datetime(2020, 12, 31)


btc_df = web.DataReader("BTC-USD", 'yahoo', start, end)
```

For this particular problem, we will retrieve data from 2014 up to the end of 2020, as represented in the table provided in *Figure 4.3*. The table provides value information about High, Low, Open, and Close for the BTC stock of the trading section. This data will be used to train the models in the current chapter:

[3]:		High	Low	Open	Close	Volume	Adj Close
Date							
2014-09-16		468.174011	452.421997	465.864014	457.334015	2.105680e+07	457.334015
2014-09-17		456.859985	413.104004	456.859985	424.440002	3.448320e+07	424.440002
2014-09-18		427.834991	384.532013	424.102997	394.795990	3.791970e+07	394.795990
2014-09-19		423.295990	389.882996	394.673004	408.903992	3.686360e+07	408.903992
2014-09-20		412.425995	393.181000	408.084991	398.821014	2.658010e+07	398.821014
...		...	...	...	...	...	...
2020-12-28		27389.111328	26207.640625	26280.822266	27084.808594	4.905674e+10	27084.808594
2020-12-29		27370.720703	25987.298828	27081.810547	27362.437500	4.526595e+10	27362.437500
2020-12-30		28937.740234	27360.089844	27360.089844	28840.953125	5.128744e+10	28840.953125
2020-12-31		29244.876953	28201.992188	28841.574219	29001.720703	4.675496e+10	29001.720703
2021-01-01		29600.626953	28803.585938	28994.009766	29374.152344	4.073030e+10	29374.152344

Figure 4.3 – Listing the data retrieved from the source (Yahoo Finance)

This data can easily be plotted by plotting one of the variables just to illustrate the continuous nature of the data:

```
btc_df['Open'].plot()
```

To illustrate a bit more about the nature of the data, we can plot an excerpt of the data:

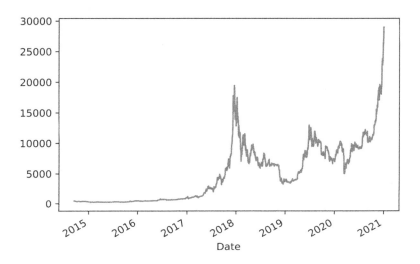

Figure 4.4 – Plot of one of the variables BTC Open retrieved from the source (Yahoo Finance)

Having defined precisely what we will be experimenting with in this section, we will move to add new models to enable us to run experiments and compare among them.

The data for the required range was conveniently saved in a file under Chapter04/ gradflow/notebooks/training_data.csv, for the period ranging from 2014 to 2020 inclusive, so it can be easily retrieved during the modeling phase.

Adding experiments

So, in this section, we will use the experiments module in **MLflow** to track the different runs of different models and post them in our workbench database so that the performance results can be compared side by side.

The experiments can actually be done by different model developers as long as they are all pointing to a shared MLflow infrastructure.

To create our first, we will pick a set of model families and evaluate our problem on each of the cases. In broader terms, the major families for classification can be tree-based models, linear models, and neural networks. By looking at the metric that performs better on each of the cases, we can then direct tuning to the best model and use it as our initial model in production.

Our choice for this section includes the following:

- **Logistic Classifier**: Part of the family of linear-based models and a commonly used baseline.

- **Xgboost**: This belongs to the family of tree boosting algorithms where many weak tree classifiers are assembled to produce a stronger model.

- **Keras**: This type of model belongs to the neural network's family and is generally indicated for situations where there is a lot of data available and relations are not linear between the features.

The steps to set up a new model are quite common and there will be overlapping and repeated code for each of the models. We will start next with a logistic regression-based classifier.

Steps for setting up a logistic-based classifier

In this sub-section, we will implement a logistic regression classifier in `scikit-learn` and train a model with our input data.

The complete notebook for this model is available in the book's repository and can be used to follow along in the `Chapter04/gradflow/notebooks/mlflow_run_logistic_regression.ipynb` file:

1. **Importing dependencies**: This section has the most important dependencies, apart from the foundational ones (pandas, NumPy, and MLflow), that we need to import the `SKLearn` model, `LogisticRegression`, and the metrics functionality, `f1_score`, that will enable us to calculate the performance:

```
import pandas
import numpy as np
import mlflow
import tensorflow
from tensorflow import keras
import mlflow.keras
from sklearn.metrics import f1_score,confusion_matrix
from sklearn.model_selection import train_test_split
```

2. **Setting up training data**: The data is read from the `training_data.csv` file:

```
pandas_df = pandas.read_csv("training_data.csv")
X=pandas_df.iloc[:,:-1]
Y=pandas_df.iloc[:,-1]
X_train, X_test, y_train, y_test = \
train_test_split(X, Y, test_size=0.33,
                        random_state=4284, stratify=Y)
```

The data is split into training and testing using the `train_test_split` function, which takes one-third of the data for testing, with the remainder being used for training.

3. **Setting up the experiment**: To set the experiment in **MLflow** programmatically, you use the `mlflow.set_experiment` method. This will create an experiment if it does not exist or associate your current run with an experiment. We use `mlflow.sklearn.autolog()` to enable the automated capabilities of MLflow to capture the metrics of our experiment:

```
mlflow.set_experiment("Baseline_Predictions")
mlflow.sklearn.autolog()
```

4. **Running the experiment**: To run your experiment, you will have to enclose it in a run using the scope keyword, `with`. The `mlflow.start_run` function is used to take care of registering your run with a specific `run_name` so that it can be identified and encloses the `fit` model, with evaluation code used to calculate the performance metrics of the `f1_score` experiment:

```
with mlflow.start_run(run_name='logistic_regression_
model_baseline') as run:
    model = LogisticRegression()
    model.fit(X_train, y_train)
    preds = model.predict(X_test)
    y_pred = np.where(preds>0.5,1,0)
    f1 = f1_score(y_test, y_pred)
    mlflow.log_metric(key="f1_experiment_score",
                        value=f1)
```

Additionally, we need to log our specific metric, `f1_experiment_score`, with the `mlflow.log_metric` function. The main reason for adding our specific method is that for each model, the autologging functionality in **MLflow** uses the default metric used by each underlying framework and generally, these metrics don't match.

After executing all the steps relating to model development, we can now navigate to our run and visualize the log of the experiment. In *Figure 4.5*, you can see the specific parameters associated with logistic regression, durations, and all the parameters used on your run:

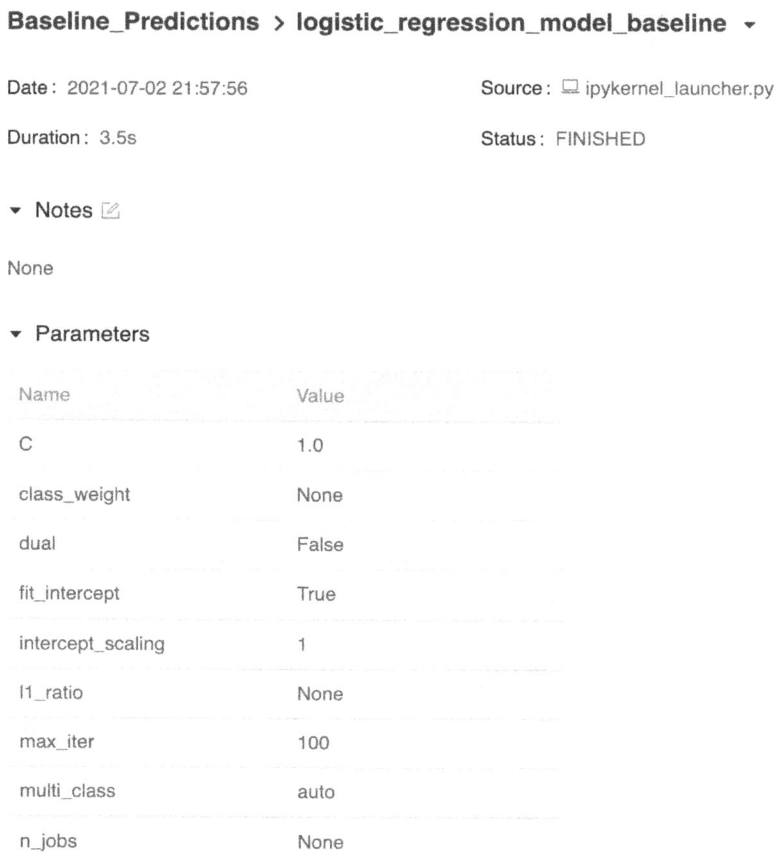

Baseline_Predictions > **logistic_regression_model_baseline** ⌄

Date : 2021-07-02 21:57:56   Source : 🖵 ipykernel_launcher.py

Duration : 3.5s   Status : FINISHED

▾ Notes ✎

None

▾ Parameters

Name	Value
C	1.0
class_weight	None
dual	False
fit_intercept	True
intercept_scaling	1
l1_ratio	None
max_iter	100
multi_class	auto
n_jobs	None

Figure 4.5 – Logistic regression model details

For SKLearn models, **MLflow** automatically logs confusion matrices and precision and recall curves that are very useful in detecting how well the model performed on training data. For instance, the *Figure 4.6* report will be stored in the artifacts of your run:

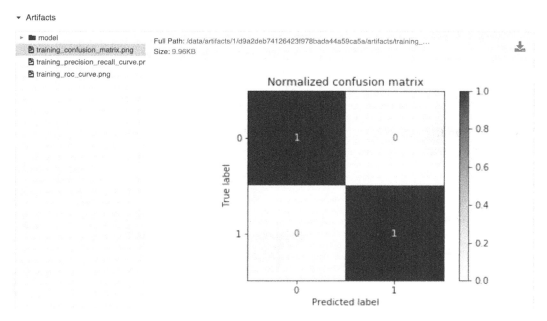

Figure 4.6 – Confusion matrix metrics

MLflow provides built-in metrics for Sklearn, providing better visibility of the model produced during training without the developer needing to produce extra code.

Steps for setting up an XGBoost-based classifier

We will now implement a gradient tree-based algorithm using the XGBoost library.

The complete notebook for this model is available in the book's repository and can be used to follow along in the Chapter04/gradflow/notebooks/mlflow_run_xgboost.ipynb file:

Importing dependencies: The XGBoost library is imported alongside the metrics function:

```
import pandas
import mlflow
import xgboost as xgb
from sklearn.metrics import f1_score
from sklearn.model_selection import train_test_split
```

1. **Retrieving data**: This step remains the same as we are splitting data and reading the data from the training_data.csv file.

2. **Setting up the experiment**: The experiment is still the same, `Baseline_`
   `Predictions`, and we need to give MLflow the instruction to automatically
   log the model through `mlflow.xgboost.autolog`:

```
mlflow.set_experiment("Baseline_Predictions")
mlflow.xgboost.autolog()
```

3. **Running the experiment**: This experiment is very similar to the previous case
   where we run the model and evaluate the metrics through `f1_score`:

```
with mlflow.start_run(
  run_name='xgboost_model_baseline') as run:
    model=xgb.train(dtrain=dtrain,params={})
    preds = model.predict(dtest)
    y_bin = [1. if y_cont > threshold else 0. for y_cont
in preds]
    f1= f1_score(y_test,y_bin)
    mlflow.log_metric(key="f1_experiment_score",
                      value=f1)
```

After executing all the steps relating to model development, we can now navigate to
our run and visualize the log of the experiment. In *Figure 4.7*, you can see the specific
parameters associated with `xgboost_model_baseline`, durations, and all the
parameters used on your run:

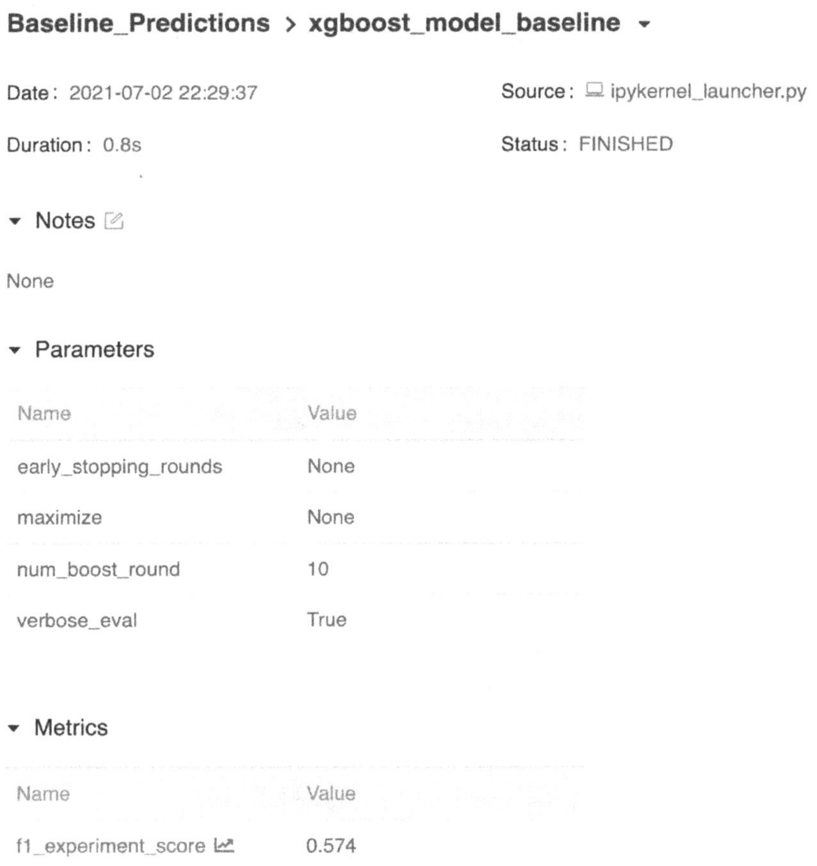

Baseline_Predictions > xgboost_model_baseline ▾

Date : 2021-07-02 22:29:37

Source : 🖵 ipykernel_launcher.py

Duration : 0.8s

Status : FINISHED

▾ Notes ✎

None

▾ Parameters

Name	Value
early_stopping_rounds	None
maximize	None
num_boost_round	10
verbose_eval	True

▾ Metrics

Name	Value
f1_experiment_score ⬈	0.574

Figure 4.7 – XGBoost classifier details in MLflow

For XGBoost models, **MLflow** automatically logs feature information and importance. We can see in *Figure 4.8* the ranking of our features in the model stored in the *Artifacts* section of the workbench:

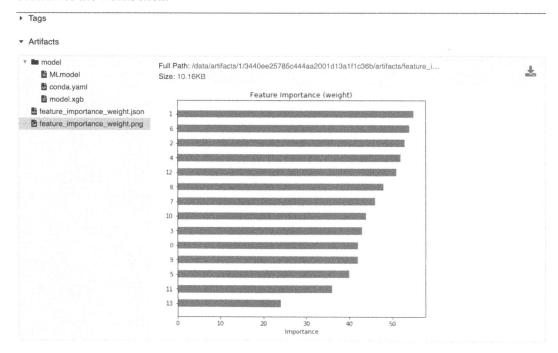

Figure 4.8 – XGBoost feature importance on MLflow

The feature importance graph in *Figure 4.8* allows the developer to have some insights into the internals of the model ascertained from the data. In this particular case, it seems that the second and seventh days of the 14 days in the input vector are the top two meaningful features. We will next implement a deep learning-based model.

Steps for setting up a deep learning-based classifier

In this section, we will implement a neural network algorithm to solve our classification problem.

The complete notebook for this model is available in the book's repository and can be used to follow along in the `Chapter04/gradflow/notebooks/mlflow_run_keras.ipynb` file:

1. **Importing dependencies**: The salient dependency in this step is `tensorflow`, as we are using it as a backend for `keras`:

```
import pandas
import numpy as np
import mlflow
import tensorflow
from tensorflow import keras
import mlflow.keras
from sklearn.metrics import f1_score,confusion_matrix
from sklearn.model_selection import train_test_split
```

2. **Retrieving data**: Refer to Step 2 in the *Steps for setting up an XGBoost-based classifier* section.

3. **Setting up the experiment**: The experiment is still the same, `Baseline_Predictions`, and we need to give MLflow the instruction to automatically log the model through `mlflow.tensorflow.autolog`:

```
mlflow.set_experiment("Baseline_Predictions")
mlflow.tensorflow.autolog()
```

Creating the model: One of the big differences compared with the neural-based model is that the creation of the model is a bit more involved than `Sklearn` or XGBoost classifiers, so we need to define the layers and architecture of the network. In this particular case, the `Sequential` architecture and the model need to be compiled as required by Tensorflow:

```
model = keras.Sequential([
  keras.layers.Dense(
    units=36,
    activation='relu',
    input_shape=(X_train.shape[-1],)
  ),
  keras.layers.BatchNormalization(),
```

```
        keras.layers.Dense(units=1, activation='sigmoid'),
    ])
model.compile(
    optimizer=keras.optimizers.Adam(lr=0.001),
    loss="binary_crossentropy",
    metrics="Accuracy"
)
```

4. **Running the model**: Running the model involves the same steps as specifying run_name and fitting the model followed by calculating the f1_score metrics:

```
with mlflow.start_run(
    run_name='keras_model_baseline') as run:
        model.fit(
            X_train,
            y_train,
            epochs=20,
            validation_split=0.05,
            shuffle=True,
            verbose=0
        )
        preds = model.predict(X_test)
        y_pred = np.where(preds>0.5,1,0)
        f1 = f1_score(y_test, y_pred)
        mlflow.log_metric(key="f1_experiment_score",
                          value=f1)
```

For keras models, **MLflow** automatically logs a myriad of neural network-related data, namely, regarding optimizers and epoch and batch sizes, as well as other relevant information that can be seen in *Figure 4.9*:

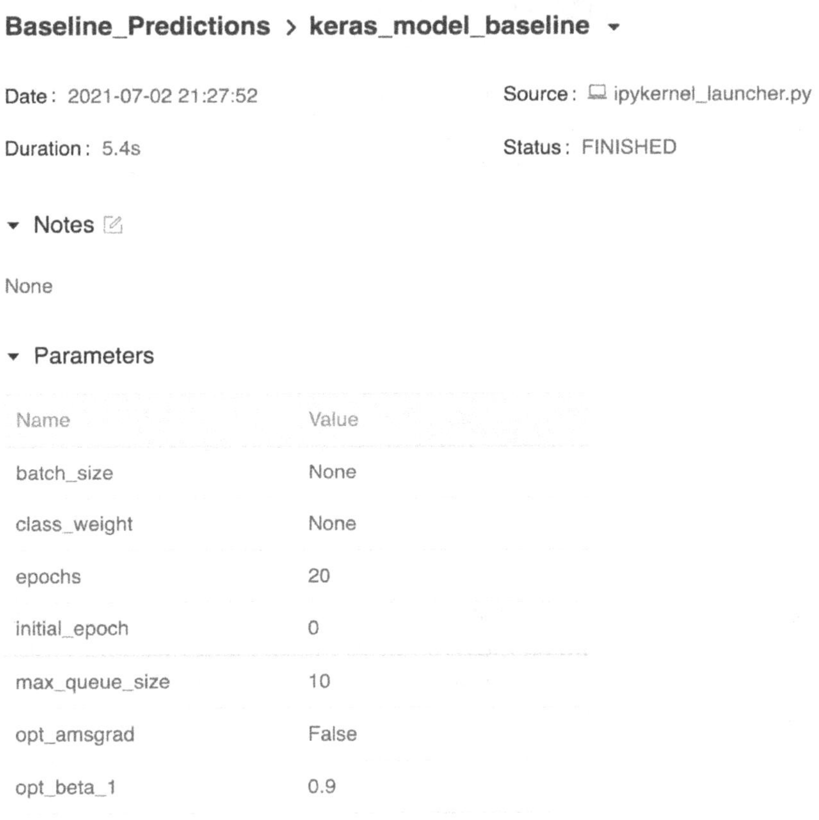

Baseline_Predictions > **keras_model_baseline** ⌄

Date : 2021-07-02 21:27:52   Source : ⌨ ipykernel_launcher.py

Duration : 5.4s   Status : FINISHED

⌄ Notes ☑

None

⌄ Parameters

Name	Value
batch_size	None
class_weight	None
epochs	20
initial_epoch	0
max_queue_size	10
opt_amsgrad	False
opt_beta_1	0.9

Figure 4.9 – Keras classifier model details

Additionally, **TensorFlow** logs can be hooked into a TensorBoard. This is a TensorFlow built-in tool to provide visualizations and metrics for the machine learning workflow. Interfaces are created so that the model developer can leverage the native TensorFlow instrumentation and specialized visualization tooling.

Having set up our classifiers in our platform, in the next section, we are ready to compare the performance of the different classifiers developed using MLflow.

Comparing different models

We have run the experiments in this section for each of the models covered and verified all the different artifacts. Just by looking at our baseline experiment table, and by selecting the common custom metric, f1_experiment_score, we can see that the best performing model is the logistic regression-based model, with an F-score of 0.66:

	Start Time	Run Name	Models	Metrics
				f1_experiment_score
☐	⊘ 2021-02-16 12:01:	keras_...	🔲 keras	0.583
☐	⊘ 2021-02-16 11:53:	logistic...	🔲 sklearn	0.663
☐	⊘ 2021-02-16 11:53:	xgboos...	🔲 xgboost	0.574
		Load more		

Figure 4.10 – Comparing different model performance in terms of the goal metric

Metrics can also be compared side by side, as shown in the excerpt in *Figure 4.11*. On the left side, we have the SKlearn model, and on the right the XGBoost model, with the custom metrics of f1_experiment_score. We can see that the metrics provided by both are different and, hence, the reason for custom metrics when we have different models:

Metrics

f1_experiment_score 📈	0.663	0.574
training_accuracy_score 📈	0.562	
training_f1_score 📈	0.52	
training_log_loss 📈	0.682	
training_precision_score 📈	0.552	
training_recall_score 📈	0.562	
training_roc_auc_score 📈	0.564	
training_score 📈	0.562	

Figure 4.11 – Metrics of the Sklearn model

After comparing the metrics, it becomes clear that the best model is logistic regression. To improve the model, in the next section, we will optimize its parameters with state-of-the-art techniques and use MLflow experiment features to achieve that.

Tuning your model with hyperparameter optimization

Machine learning models have many parameters that allow the developer to improve performance and control the model that they are using, providing leverage to better fit the data and production use cases. Hyperparameter optimization is the systematic and automated process of identifying the optimal parameters for your machine learning model and is critical for the successful deployment of such a system.

In the previous section, we identified the best family (in other words, LogisticRegression) model for our problem, so now it's time to identify the right parameters for our model with MLflow. You can follow along in the following notebook in the project repository, Chapter04/gradflow/notebooks/hyperopt_optimization_logistic_regression_mlflow.ipynb:

1. **Importing dependencies**: We will use the hyperopt library, which contains multiple algorithms to help us carry out model tuning:

    ```
    from hyperopt import tpe
    from hyperopt import STATUS_OK
    from hyperopt import Trials
    from hyperopt import hp
    from hyperopt import fmin
    from sklearn.linear_model import LogisticRegression
    from sklearn.model_selection import cross_val_score
    from sklearn.model_selection import train_test_split
    ```

2. **Defining an objective function**: The objective function is the most important step
    of the process, essentially defining what we want to achieve with our optimization.
    In our particular case, we want to optimize the f1_score metric in our model.
    The way optimization works in hyperopt is through minimization, but in our
    case, we want the maximum possible f1_score metric. So, the way we define our
    loss (the function to minimize) is as the inverse of our f1_score metric, as in
    loss = 1-fscore, so the minimization of this function will represent the best
    f1_score metric. For each run of the model's parameters, we will enclose it in
    an mlflow.start_run(nested=True) in such a way that each optimization
    iteration will be logged as a sub run of the main job, providing multiple advantages
    in terms of comparing metrics across runs:

```
N_FOLDS = 3
MAX_EVALS = 10


def objective(params, n_folds = N_FOLDS):
    # Perform n_fold cross validation with
    #hyperparameters
    # Use early stopping and evaluate based on ROC AUC
    mlflow.sklearn.autolog()
    with mlflow.start_run(nested=True):
        clf = LogisticRegression(**params,
                                 random_state=0,
                                 verbose =0)
        scores = cross_val_score(clf, X_train,
                                 y_train, cv=5,
                                 scoring='f1_macro')

        # Extract the best score
        best_score = max(scores)
```

```
# Loss must be minimized
loss = 1 - best_score

# Dictionary with information for evaluation
return {'loss': loss, 'params': params,
        'status': STATUS_OK}
```

3. **Running optimization trials**: The trials step allows us to run multiple experiments with the logistic regression algorithm and help us identify the best possible configuration for our model, which will be stored under the `best` variable. The core function is the minimization represented by `fmin(fn = objective, space = space, algo = tpe.suggest, max_evals = MAX_EVALS, trials = bayes_trials)`, where we provide the parameter space and objective function as previously defined:

```
# Algorithm
tpe_algorithm = tpe.suggest

# Trials object to track progress
bayes_trials = Trials()
mlflow.set_experiment("Bayesian_param_tuning")
with mlflow.start_run():
    best = fmin(fn = objective, space = space,
                algo = tpe.suggest,
                max_evals = MAX_EVALS,
                trials = bayes_trials)
```

4.   After running the experiment for a few minutes, we can now review the experiments in **MLflow**. *Figure 4.12* represents the experiment and the nested hierarchy of the multiple experiments run under the umbrella of the `Hyperopt_Optimization` experiment:

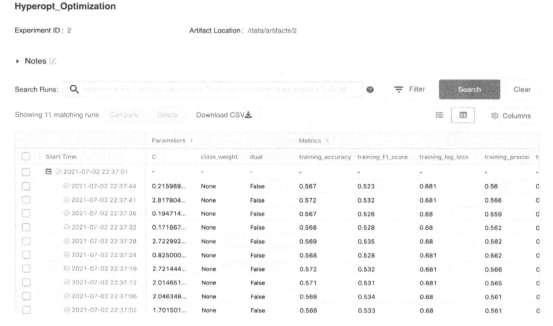

Hyperopt_Optimization

Experiment ID : 2   Artifact Location : /data/artifacts/2

▸ Notes

Start Time	Parameters >			Metrics <			
	C	class_weight	dual	training_accuracy	training_f1_score	training_log_loss	training_precisi
⊟ ⊘ 2021-07-02 22:37:01	-	-	-	-	-	-	-
⊘ 2021-07-02 22:37:44	0.215989...	None	False	0.567	0.523	0.681	0.56
⊘ 2021-07-02 22:37:41	2.817804...	None	False	0.572	0.532	0.681	0.566
⊘ 2021-07-02 22:37:36	0.194714...	None	False	0.567	0.526	0.68	0.559
⊘ 2021-07-02 22:37:32	0.171667...	None	False	0.568	0.528	0.68	0.562
⊘ 2021-07-02 22:37:28	2.722992...	None	False	0.569	0.535	0.68	0.562
⊘ 2021-07-02 22:37:24	0.825000...	None	False	0.568	0.528	0.681	0.562
⊘ 2021-07-02 22:37:19	2.721444...	None	False	0.572	0.532	0.681	0.566
⊘ 2021-07-02 22:37:12	2.014651...	None	False	0.571	0.531	0.681	0.565
⊘ 2021-07-02 22:37:06	2.046348...	None	False	0.568	0.534	0.68	0.561
⊘ 2021-07-02 22:37:02	1.701501...	None	False	0.568	0.533	0.68	0.561

Figure 4.12 – Listing all the nested runs of the hyperparameter tuning

5.   By clicking on the **compare** option, we have the results displayed in *Figure 4.13*. You can analyze multiple runs of the optimization of parameters in sequence, reviewing the implication of specific parameters in relation to performance metrics such as `training_f1_score` and the solver:

Parameters

C	2.722992654722224	2.0463482360417893	1.7015010837141984	2.817804606149807
class_weight	None	None	None	None
dual	False	False	False	False
fit_intercept	True	True	True	False
intercept_scaling	1	1	1	1
l1_ratio	None	None	None	None
max_iter	512	969	538	641
multi_class	auto	auto	auto	auto
n_jobs	None	None	None	None
penalty	l2	l2	l2	l2
random_state	0	0	0	0
solver	sag	saga	sag	lbfgs
tol	2.1621440749472694e-05	2.21729551904932778e-05	3.247189008512365e-05	2.18879038427147793e-05
verbose	0	0	0	0
warm_start	True	False	True	True

Metrics

training_accuracy_score	0.569	0.568	0.568	0.572

Figure 4.13 – Listing all the nested runs of the hyperparameter tuning

6.   We can easily compare in the same interface the different solvers and implications for our performance metric, providing further insights into our modeling phase:

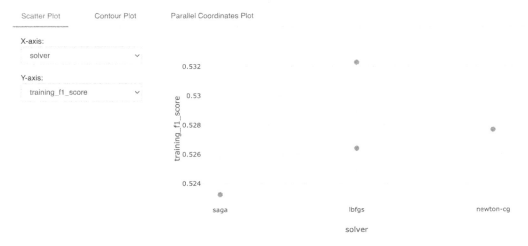

Figure 4.14 – Listing all the nested runs of the hyperparameter tuning

We concluded this section by optimizing the parameters of the most performant model for our current problem. In the next chapter of the book, we will be using the information provided by the best model to delve into the life cycle of the model management in **MLflow**.

Summary

In this chapter, we introduced the experiments component of MLflow. We got to understand the logging metrics and artifacts in MLflow. We detailed the steps to track experiments in MLflow.

In the final sections, we explored the use case of hyperparameter optimization using the concepts learned in the chapter.

In the next chapter, we will focus on managing models with MLflow using the models developed in this chapter.

Further reading

To consolidate your knowledge further, you can consult the documentation available at the following links:

- `https://www.mlflow.org/docs/latest/tracking.html`
- `https://en.wikipedia.org/wiki/Hyperparameter_optimization`

5
Managing Models with MLflow

In this chapter, you will learn about different features for model management in MLflow. You will learn about the model life cycle in MLflow and we will explain how to integrate it with your regular development workflow and how to create custom models not available in MLflow. A model life cycle will be introduced alongside the Model Registry feature of MLflow.

Specifically, we will look at the following sections in this chapter:

- Understanding models in MLflow
- Exploring model flavors in MLflow
- Managing models and signature schemas
- Managing the life cycle with a model registry

From a workbench perspective, we would like to use MLflow to manage our models and implement a clear model life cycle. The addition of managed model features to our benchmark leveraging MLflow will step up the quality and operations of our **machine learning engineering** solution.

Technical requirements

For this chapter, you will need the following:

- The latest version of Docker installed on your machine. If you don't already have it installed, please follow the instructions at `https://docs.docker.com/get-docker/`.

- The latest version of `docker-compose` installed. Please follow the instructions at `https://docs.docker.com/compose/install/`.

- Access to Git in the command line and installed as described at `https://git-scm.com/book/en/v2/Getting-Started-Installing-Git`.

- Access to a Bash terminal (Linux or Windows).

- Access to a browser.

- Python 3.5+ installed.

- The latest version of your machine learning workbench installed locally, described in *Chapter 3*, *Your Data Science Workbench*.

Understanding models in MLflow

On the MLflow platform, you have two main components available to manage models:

- **Models**: This module manages the format, library, and standards enforcement module on the platform. It supports a variety of the most used machine learning models: sklearn, XGBoost, TensorFlow, H20, fastai, and others. It has features to manage output and input schemas of models and to facilitate deployment.

- **Model Registry**: This module handles a model life cycle, from registering and tagging model metadata so it can be retrieved by relevant systems. It supports models in different states, for instance, live development, testing, and production.

An MLflow model is at its core a packaging format for models. The main goal of MLflow model packaging is to decouple the model type from the environment that executes the model. A good analogy of an MLflow model is that it's a bit like a **Dockerfile** for a model, where you describe metadata of the model, and deployment tools upstream are able to interact with the model based on the specification.

As can be seen in the diagram in *Figure 5.1*, on one side you have your model library, for instance, TensorFlow or sklearn. At the core of MLflow, you have the MLflow model format, which is able to be served in a multitude of flavors (model formats) to cater to different types of inference tools on-premises and in the cloud:

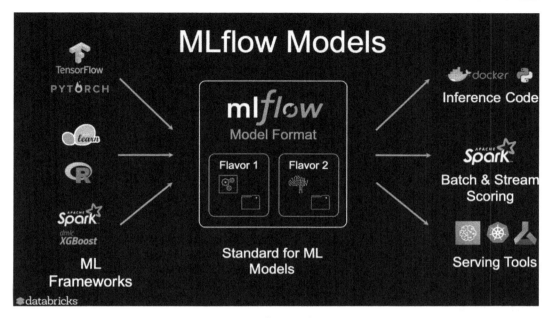

Figure 5.1 – MLflow models diagram

Figure 5.1 was extracted from the URL `https://www.infoq.com/presentations/mlflow-databricks/#`.

The central piece of the definition of MLflow models is the MLflow model file, as depicted in the next screenshot:

```
artifact_path: model
flavors:
  python_function:
    env: conda.yaml
    loader_module: mlflow.sklearn
    model_path: model.pkl
    python_version: 3.7.6
  sklearn:
    pickled_model: model.pkl
    serialization_format: cloudpickle
    sklearn_version: 0.22.2.post1
run_id: 75c2c826870d4d6082b3c6e10934a99f
signature:
  inputs: '[{"type": "double"}, {"type": "double"}, {"type": "double"}, {"type": "double"},
    {"type": "double"}, {"type": "double"}, {"type": "double"}, {"type": "double"}]'
  outputs: '[{"type": "long"}]'
utc_time_created: '2021-02-22 15:47:40.557303'
```

Figure 5.2 – An example of an MLmodel file

An MLmodel example can be seen in *Figure 5.2* and provides the following information:

- **run_id**: This is a reference to the run of the model of the project that allowed the creation of the model.

- **time_created**: The timestamp of when the model was created.

- **flavors**: Flavors are different types of models, whether that is the native models (TensorFlow, Keras, sklearn, and so on) supported by MLflow or the pyfunc model provided by MLflow.

- **signature**: This is the component of the MLmodel that defines the model signature and allows you to, in some way, type the inference process of your model. It allows the validation of input data that needs to match the signature of the model.

The **MLflow Models** module provides you with the ability to deploy your models in a native environment of the library of your model or in a generic interoperable MLflow environment called pyfunc. This function is supported in any environment that supports Python, providing flexibility to the deployer of the model on how best to run the model once logged in MLflow:

1. In the GitHub repo of the project, please go to the Gradflow folder and start the environment in this chapter by running the following command:

```
make
```

2. You can run all the cells including the model cell depicted in *Figure 5.3*:

```
mlflow.set_experiment("Baseline_Predictions_Mlflow_Check")
mlflow.tensorflow.autolog()

model = keras.Sequential([
  keras.layers.Dense(
    units=36,
    activation='relu',
    input_shape=(X_train.shape[-1],)
  ),
  keras.layers.BatchNormalization(),
  keras.layers.Dense(units=1, activation='sigmoid'),
])

model.compile(
  optimizer=keras.optimizers.Adam(lr=0.001),
  loss="binary_crossentropy",
  metrics="Accuracy"
)
with mlflow.start_run(run_name='keras_model_baseline') as run:
    model.fit(X_train, y_train, epochs=20, validation_split=0.05, shuffle=True,verbose=)
```

Figure 5.3 – An example of an MLmodel file

The model in *Figure 5.3* should be very similar to the one used in *Chapter 4, Experiment Management in MLflow*. Using `mlflow.start_run`, you can start logging your model in MLflow and use the innate capabilities of the platform to capture relevant details of the model being developed.

3. You can now explore the `MLmodel` file in MLflow:

Figure 5.4 – An example of an MLmodel file

4. Explore the `conda` file in MLflow:

Figure 5.5 – An example of an MLmodel file

5. Load the model as `MLflow Pyfunc` for prediction:

```
import mlflow
logged_model = '/data/
artifacts/1/132e6fa332f2412d85f3cb9e6d6bc933/artifacts/
model'


# Load model as a PyFuncModel.
loaded_model = mlflow.pyfunc.load_model(logged_model)


# Predict on a Pandas DataFrame.
import pandas as pd
loaded_model.predict(pd.DataFrame(X_test))
```

Alternatively, the model can be loaded in the native H5 Keras format and loaded to a completely different application, as shown in *Figure 5.4*, by using the `/data/model/model.h5 file`.

After introducing in this section the concept of models in MLflow, we will next delve a bit deeper into the different types of models in MLflow.

Exploring model flavors in MLflow

Model flavors in MLflow are basically the different models of different libraries supported by MLflow. This functionality allows MLflow to handle the model types with native libraries of each specific model and support some of the native functionalities of the models. The following list presents a selection of representative models to describe and illustrate the support available in MLflow:

*   `mlflow.tensorflow`: TensorFlow is by far one of the most used libraries, particularly geared toward deep learning. MLflow integrates natively with the model format and the monitoring abilities by saving logs in TensorBoard formats. Auto-logging is supported in MLflow for TensorFlow models. The Keras model in *Figure 5.5* is a good example of TensorFlow support in MLflow.

- `mlflow.h2o`: H2O is a complete machine learning platform geared toward the automation of models and with some overlapping features with MLflow. MLflow provides the ability to load (`load_model`) and log models (`log_model`) in H2O native format, allowing interoperability between the tools. Unfortunately, as of the current MLflow version, you can't use auto-logging on h2o models:

```
mlflow.h2o.load_model(...)
mlflow.h2o.log_model(...)
```

- `mlflow.spark`: MLflow integrates with the Apache Spark library natively through two main interfaces: Spark MLlib for machine learning and the MLeap platform (`https://combust.github.io/mleap-docs/`). Mleap is more of a deployment platform while MLlib is more of a library that you can add to your projects.

A very comprehensive list of flavors/formats is supported by MLflow and their usage and support can be read about here: `https://www.mlflow.org/docs/latest/python_api/index.html`.

Custom models

We can delve into the next excerpt of code and the custom `RandomPredictor` model. As long as you provide a class with an interface with the `fit` and `predict methods`, you can have your own custom MLflow model:

```
class RandomPredictor(mlflow.pyfunc.PythonModel):
  def __init__(self):
    pass

  def fit(self):
    pass

  def predict(self, context, model_input):
    return model_input.apply(
        lambda column: random.randint(0,1))
```

In the preceding `class`, we basically use a random probability, and it can be used as a sample model in a system where you want to make sure that your model is better than a random model.

In this section, we introduced different types of model flavors and the creation of a custom mode. We will next look at some of the schemas and signature features of MLflow.

Managing model signatures and schemas

An important feature of MLflow is to provide an abstraction for input and output schemas of models and the ability to validate model data during prediction and training.

MLflow throws an error if your input does not match the schema and signature of the model during prediction:

1. We will next look at a code listing of a simple model of digit classification (the details of the dataset are available here: https://archive.ics.uci.edu/ml/datasets/Optical+Recognition+of+Handwritten+Digits). The following code flattens the image into a pandas DataFrame and fits a model to the dataset:

```
from sklearn import datasets, svm, metrics
from sklearn.model_selection import import train_test_split
import mlflow


digits = datasets.load_digits()
n_samples = len(digits.images)
data = digits.images.reshape((n_samples, -1))


clf = svm.SVC(gamma=0.001)


X_train, X_test, y_train, y_test = train_test_split(
    data, digits.target, test_size=0.5, shuffle=False)
mlflow.sklearn.autolog()


with mlflow.start_run():
    clf.fit(X_train, y_train)
```

2. We'll look at the previous code listing, which you can run in a new notebook and navigate through the MLflow UI to investigate in more depth the MLmodel generated in *Figure 5.6*:

```
artifact_path: model
flavors:
  python_function:
    env: conda.yaml
    loader_module: mlflow.sklearn
    model_path: model.pkl
    python_version: 3.7.6
  sklearn:
    pickled_model: model.pkl
    serialization_format: cloudpickle
    sklearn_version: 0.22.2.post1
run_id: 57d4216eeea1499c8607b1d3f6265775
signature:
  inputs: '[{"type": "double"}, {"type": "double"}, {"type": "double"}, {"type": "double"},
    {"type": "double"}, {"type": "double"}, {"type": "double"}, {"type": "double"},
    {"type": "double"}, {"type": "double"}, {"type": "double"}, {"type": "double"},
    {"type": "double"}, {"type": "double"}, {"type": "double"}, {"type": "double"},
    {"type": "double"}, {"type": "double"}, {"type": "double"}, {"type": "double"},
    {"type": "double"}, {"type": "double"}, {"type": "double"}, {"type": "double"},
    {"type": "double"}, {"type": "double"}, {"type": "double"}, {"type": "double"}]'
  outputs: '[{"type": "long"}]'
utc_time_created: '2021-03-11 19:28:54.202276'
```

Figure 5.6 – Sample of an MLmodel file

3. The MLmodel file contains the signature in JSON of input and output files. For some of the flavors autologged, we will not be able to infer the signature automatically so you can provide the signature inline when logging the model:

```
# flatten the images
from mlflow.models.signature import infer_signature
with mlflow.start_run(run_name='untuned_random_forest'):
    ...
    signature = infer_signature(X_train,
        wrappedModel.predict(None, X_train))
    mlflow.pyfunc.log_model("random_forest_model",
                            python_model=wrappedModel,
                            signature=signature)
```

In the previous code block, the signature of the model is provided by the `infer_signature` method. As the model is logged through `log_model`, the signature is provided. One important advantage of the signatures being logged alongside the model is that they can serve as documentation and metadata for the model. Third-party systems can consume the metadata and interact with the models by validating the data or generating documentation for the models.

In this section, we introduced the model schema and signature features of MLflow models. We will now move on to the other critical module in this space, namely the Model Registry.

Introducing Model Registry

MLflow Model Registry is a module in MLflow that comprises a centralized store for Models, an API allowing the management of the life cycle of a model in a registry.

A typical workflow for a machine learning model developer is to acquire training data; clean, process, and train models; and from there on, hand over to a system or person that deploys the models. In very small settings, where you have one person responsible for this function, it is quite trivial. Challenges and friction start to arise when the variety and quantity of models in a team start to scale. A selection of common friction points raised by machine learning developers with regards to storing and retrieving models follows:

- Collaboration in larger teams
- Phasing out stale models in production
- The provenance of a model
- A lack of documentation for models
- Identifying the correct version of a model
- How to integrate the model with deployment tools

The main idea behind **MLflow Model Registry** is to provide a central store model in an organization where all the relevant models are stored and can be accessed by humans and systems. A good analogy would be a Git repository for models with associated relevant metadata and centralized state management for models.

In the MLflow UI (available in your local environment), you should click on the tab on the right side of **Experiments** with the label **Models** as indicated by the arrow:

1. Through this module, you are able to list all the models registered, search by name, or create by name. For each model, you can see the label of the latest version and the specific versions that are in staging or production:

Figure 5.7 – Model Registry UI

2. A new model can be created by clicking on the **Create Model** button where a relevant name can be given to a specific model as shown in *Figure 5.8*:

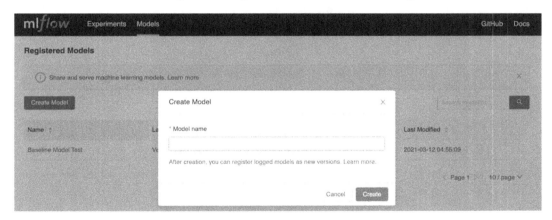

Figure 5.8 – Model Registry UI – Create Model

3. You can also create models in MLflow by running into the **Experiments** model and choosing one of your models, and from there, specifically deciding to register the model. You will have to associate your run with an existing model or create a new model name to associate with this particular type of model thereafter:

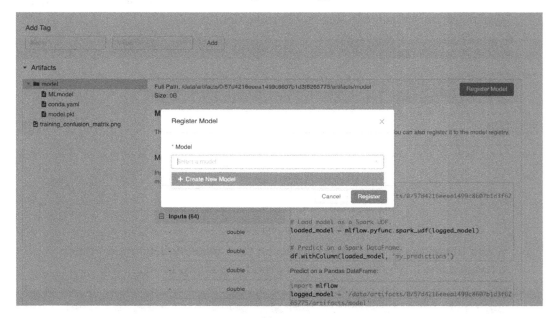

Figure 5.9 – Model Tracking UI – Create New Model

When you add a new model, MLflow automatically increases the version and labels this version as the latest version and everyone in the organization can query the registry for the latest version of a model for a given problem.

Adding your best model to Model Registry

Everything that can be done in the UI in MLflow can also be implemented through the MLflow API.

We can quickly go back to our use case of stock market prediction and add our first baseline model to Model Registry and run the `hyperopt_optimization_logistic_regression_mlflow.ipynb` notebook, available in the repo of this chapter, and sort the runs according to the F1 score metrics in descending order as represented by *Figure 5.10*:

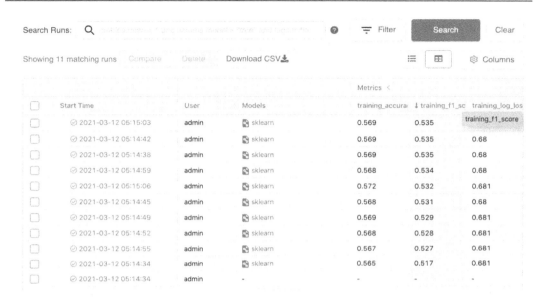

Figure 5.10 – Selecting the best model

From there, you should be able to register the best model with the name `BTC StockPrediction` as represented in *Figure 5.11*:

Figure 5.11 – Naming your model

By returning to the models module, you will notice, as represented in *Figure 5.12*, your newly created model under **Version 1**:

Figure 5.12 – Registered Models

Having introduced the functionalities of Model Registry, in the next section, we will describe a model development life cycle to help organize the management of your models.

Managing the model development life cycle

Managing the model life cycle is quite important when working in a team of more than one model developer. It's quite usual for multiple model developers to try different models within the same project, and having a reviewer decide on the model that ends up going to production is quite important:

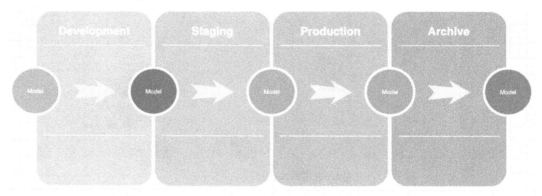

Figure 5.13 – Example of a model development life cycle

A model in its life cycle can undergo the following stages if using a life cycle similar to the one represented in *Figure 5.13*:

- **Development**: The state where the model developer is still exploring and trying out different approaches and is still trying to find a reasonable solution to their machine learning problem.

- **Staging**: The state where the model can be tested automatically with production-type traffic.

- **Production**: When the model is ready to handle real-life production traffic.

- **Archive**: When the model no longer serves the business purpose that it was initially developed for, it will be archived and its metadata stored for future reference or compliance.

For instance, a reviewer or supervisor, as represented in *Figure 5.14*, can move a model from the **Development** state to **Staging** for further deployment in a test environment and the model can be transitioned into production if approved by reviewers:

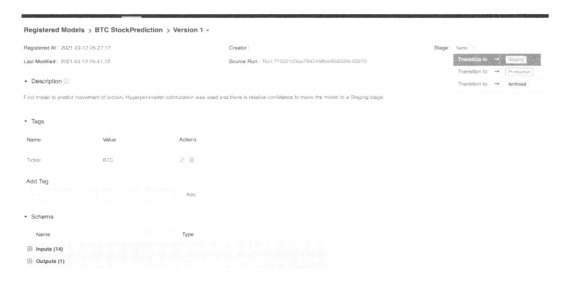

Figure 5.14 – Example of a model development life cycle

When transitioning from a state in MLflow, you have the option to send the model in an existing state to the next state:

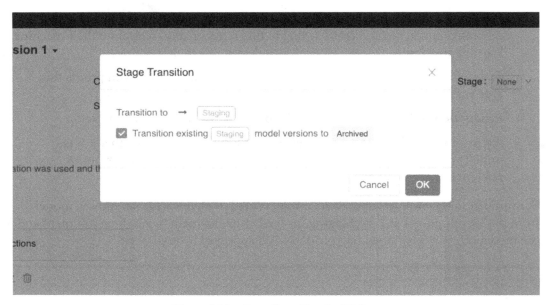

Figure 5.15 – Stage Transition in MLflow

The transitions from the **Staging** to **Production** stages in a mature environment are meant to be done automatically, as we will demonstrate in the upcoming chapters of the book.

With this section, we have concluded the description of the features related to models in MLflow.

Summary

In this chapter, we first introduced the Models module in MLflow and the support for different algorithms, from tree-based to linear to neural. We were exposed to the support in terms of the logging and metrics of models and the creation of custom metrics.

In the last two sections, we introduced the Model Registry model and how to use it to implement a model life cycle to manage our models.

In the next chapters and section of the book, we will focus on applying the concepts learned so far in terms of real-life systems and we will architect a machine learning system for production environments.

Further reading

In order to solidify your knowledge and dive deeper into the concepts introduced in this chapter, you should look at the following links:

- `https://www.mlflow.org/docs/latest/models.html`

- `https://www.mlflow.org/docs/latest/model-registry.html`

- `https://www.slideshare.net/Hadoop_Summit/introducing-mlflow-an-open-source-platform-for-the-machine-learning-life cycle-for-onprem-or-in-the-cloud`

Section 3: Machine Learning in Production

This section will cover the different features available in MLflow for reliably taking a model from the development phase to a production environment seamlessly.

The following chapters are covered in this section:

- *Chapter 6, Introducing ML Systems Architecture*
- *Chapter 7, Data and Feature Management*
- *Chapter 8, Training Models with MLflow*
- *Chapter 9, Deployment and Inference with MLflow*

6
Introducing ML Systems Architecture

In this chapter, you will learn about general principles of **Machine Learning** (**ML**) systems architecture in the broader context of **Software Engineering** (**SWE**) and common issues with deploying models in production in a reliable way. You will also have the opportunity to follow along with architecting our ML systems. We will briefly look at how with MLflow, in conjunction with other relevant tools, we can build reliable and scalable ML platforms.

Specifically, we will look at the following sections in this chapter:

- Understanding challenges with ML systems and projects
- Surveying state-of-the-art ML platforms
- Architecting the PsyStock ML platform

You will follow a process of understanding the problem, studying different solutions from lead companies in the industry, and then developing your own relevant architecture. This three-step approach is transferrable to any future ML system that you want to develop.

Technical requirements

For this chapter, you will need to meet the following prerequisites:

- The latest version of Docker installed on your machine. If you don't already have it installed, please follow the instructions at `https://docs.docker.com/get-docker/`.

- The latest version of `docker-compose` installed. Please follow the instructions at `https://docs.docker.com/compose/install/`.

- Access to Git in the command line and installed as described at `https://git-scm.com/book/en/v2/Getting-Started-Installing-Git`.

- Access to a Bash terminal (Linux or Windows).

- Access to a browser.

- Python 3.5+ installed.

- The latest version of your ML platform installed locally as described in *Chapter 3, Your Data Science Workbench*.

Understanding challenges with ML systems and projects

Implementing a product leveraging ML can be a laborious task as some new concepts need to be introduced in the book around best practices of ML systems architecture.

So far in this book, we have shown how MLflow can enable the everyday model developer to have a platform to manage the ML life cycle from iteration on model development up to storing their models on the model registry.

In summary, at this stage, we have managed to create a platform for the model developer to craft their models and publish the models in a central repository. This is the ideal stage to start unlocking potential in the business value of the models created. In an ML system, to make the leap from model development to a model in production, a change of mindset and approach is needed. After unlocking the value and crafting models, the exploitation phase begins, which is where having an ML systems architecture can set the tone of the deployments and operations of your models.

ML systems are a specialization of the traditional SWE area and so we can and should leverage the body of knowledge in the SWE realm to architect our systems. Relevant concepts in SWE to our context are the following:

- **Separation of concerns**: A complete system should be separated into different components. Each of the components of the system should be independent and focused on doing one thing well. For instance, a training component should be specialized in training and not doing scoring at the same time.

- **Autonomy**: Each component of the system should stand as an independent autonomous unit and be deployable independently. For example, the deployment of your API system should be independent of the deployment of the training system.

- **Resilience**: One consequence of the separation of concerns and modularity is that we must make sure that if one component of the wider system is faulty, this doesn't affect independent components. If a batch scoring mechanism of a machine platform is broken, it shouldn't affect the real-time system.

- **Scalability**: We should be able to scale the different components of our system independently and in accordance with its workload.

- **Testability**: This represents the ability of a system to being tested and its functionality being verified against a set of representative inputs. In ML systems, this is particularly hard given the non-deterministic nature of models.

- **Continuous deployment/delivery**: This represents the ability to deploy systems in shorter cycles with almost no friction between a change in the code, configuration, models, or data, in the ML case, to have the new version of the system.

- **Composability**: We should be able to reuse the components of our systems in future projects to increase the return on investment. So, an ML engineer needs to be sure that the code and components being developed are easily reusable and/or interoperable with other systems.

- **Maintainability**: This is the ease at which a system can be modified, fixed, and improved to meet and adapt to the demands of a changing environment.

At this stage, we can briefly introduce and refine our use case, of stock prediction, to develop our ML platform in the PsyStock company.

Based on the work done so far in prototyping models to **predict the price of Bitcoin**, the business development department of the company decided to start its first product as a **Prediction API for cryptocurrencies** as they are becoming a popular technology in the corporate world. A team was assembled that decided to investigate challenges and state-of-the-art platforms, and then architect the company's own platform.

An ML project generally involves many departments of a company. Imagining the hypothetical case of PsyStock, a typical ML project team involves the following stakeholders:

- **Data science team**: Responsible for building and developing the model with the goal of achieving the highest accuracy on their prediction of cryptocurrency prices and market movements.

- **ML/data engineering team**: Responsible for the engineering components, including data acquisition, preparation, training, and deployment, and is interested in the system being correctly deployed and running on spec in production.

- **Infrastructure team**: Responsible for providing compute and infrastructure resources. Expects that the system will not cause an operational load to the team.

- **Product team**: Provides integration with the web platforms and the overall software of the company and drives the feature creation, ensuring a speedy inference speed.

- **Business development/marketing team**: Packages and markets the product and monitors the business performance of the product.

In the next section, we will understand general challenges in ML systems and projects.

ML is an important application of technology to help unlock value in organizations using data. In the application of ML in the business world, there are no standard practices defined and a big number of organizations struggle to get products backed by ML in production.

In the real world, a naive way to move models into production would consist of the following steps:

1. Data scientist produces a model in a notebook environment and implements the code in R.

2. The data scientist shares the notebook with the engineering team, signaling that they're ready to send their model to production.

3. The engineering team reimplements the training process in a language that they can understand, in this case, Python.

4. A long process of trial and error until the data science team and engineering team are in agreement that the model produced by the bespoke training system is producing equivalent outputs to the original one.

5. A new system is created and developed to score systems and the engineering team notes a high latency. The model is sent to redevelopment as it can't be redeveloped in the current status.

The situation described in the previous paragraph is more common than you might imagine. It is described in detail in the paper by *D. Sculley et al., Hidden Technical Debt in Machine Learning Systems (2015)*. The following risk factors and technical debt related to implementing ML platforms naively were identified by a team at Google:

- **Boundary erosion**: ML systems by their nature mix signals of different logical domains. Maintaining clear logic of business domains as is possible in SWE is challenging. Another natural issue is the temptation of using a model output as input of a third model, *A*, and changes might have unexpected effects in model *B*.

- **Costly data dependencies**: Fresh, accurate, and dependable data is the most important ingredient of an ML system. For example, in the cryptocurrency case, in order to be able to predict, an external API might be consulted in combination with social network sentiment signals. At a given point, one of the data signals might be unavailable, making one of the components unavailable. Data distributions in the real world can change, causing the model inference to be irrelevant in the real world.

- **Feedback loops**: In some contexts, the model influences the data selected for training. Credit scoring is a good example of such a case. A model that decides who gets credit for the next re-training of the model will select training data from the population that was affected by the model. Analyzing the effect of the model on the ground data is important to take into consideration when developing your model.

- **System-level anti patterns**: ML systems are notoriously known for harboring glue code with different packages and without proper abstractions. In some cases, multiple languages are used to implement in the library given the iterative nature of developing code in a notebook.

- **Configuration management**: Generally left as an afterthought in ML systems, information about the configuration that yielded a particular result is paramount for the post-analysis of models and deployment. Not using established configuration management practices can introduce errors in ML pipelines.

- **Monitoring and testing**: Integration testing and unit testing are common patterns in SWE projects that due to the stochastic nature of ML projects are harder to implement.

One important practice to tackle the challenges in ML systems is to add extensive tests on critical parts of the process, on your code, during model training and when running on your system, as shown in *Figure 6.1*:

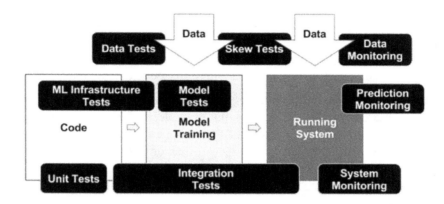

Figure 6.1 – Testing in ML systems extracted from https://research.google/pubs/pub46555/

What *Figure 6.1* illustrates is one approach to address technical debt by testing the different parts of the system through standard software practices with the addition of specialized monitoring for data predictions. The important new additions are tests on data and tests on the model, so testing incoming data and training data and at the same time being able to monitor these tests and decide whether the system passes the relevant criteria is critical.

MLflow as a platform addresses some of the issues referred to in this section as problems for ML systems. MLflow is focused on a specific set of dimensions of the ML technical debt and is a good pillar component to create an ML platform.

In the next section, we will look at some examples of state-of-the-art robust ML engineering systems, to guide our development.

Surveying state-of-the-art ML platforms

At a high level, a mature ML system has the components outlined in *Figure 6.2*. These components are ideally independent and responsible for one particular feature of the system:

Figure 6.2 – Components of an ML platform

Following the lead from SWE modularization, these general components allow us to compare different ML platforms and also specify our PsyStock requirements for each of the components. The components that we choose to use as a reference for architecture comparison are the following:

- **Data and feature management**: The component of data and feature management is responsible for data acquisition, feature generation, storing, and serving the modules upstream.

- **Training infrastructure**: The component that handles the process of the training of models, scheduling, consuming features, and producing a final model.

- **Deployment and inference**: The responsibility of this unit is for the deployment inference and batch scoring of a model. It is the external face of the system and is accessible to external systems.

- **Performance and monitoring**: A component that handles observability, metrics posted by different systems, and monitoring systems in production.

- **Model management**: Manages model artifact versions and the life cycle of models.

- **Workflow management**: The component responsible for orchestrating batch workflows and processing pipelines.

After having described the different components of an ML platform, we will look at some examples starting with Uber's Michelangelo.

Getting to know Michelangelo

Uber was one of the first companies to document widely their realization that an ML platform was critical to unlocking value on the data produced.

The internal motivations at Uber to build the platform were the following:

- Limited impact of ML due to huge resources needed when translating a local model into production.

- Unreliable ML and data pipelines.

- Engineering teams had to create custom serving containers and systems for the systems at hand.

- Inability to scale ML projects.

The following *Figure 6.3* (retrieved from `https://eng.uber.com/michelangelo-machine-learning-platform`) shows the different components of Michelangelo. One significant component is the data component of the Uber infrastructure decoupling real-time data infrastructure with streaming systems such as Kafka to acquire data from the outside from where the data flows to a training process, and from there to scoring in both real-time and offline mode. Distinctive features are a separation of the batch world and real-time world and the existence of generic prediction services for API and batch systems:

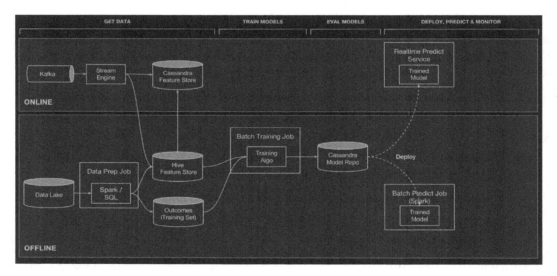

Figure.6.3 – Michelangelo architecture

The components that we choose to use as a reference for architecture comparison are the following:

- **Data and feature management**: It consists of a centralized data store with all the features that are needed to serve models and train models. The feature data store can be accessed in real time and in batch. For the batch scenario, they use a database system called Hive and for real time, they use Cassandra.

- **Training infrastructure**: Distributed training infrastructure with a tool called **Horovod** (`https://github.com/horovod/horovod`) with specialized and bespoke components and enhanced reporting. It provides custom metrics for each type of model (deep learning, models, feature importance, and so on). The output of the training job is the model repo using as a backend the Cassandra database.

- **Deployment and inference**: The systems deployed through standard SWE practices (CI/CD, rollbacks on metrics monitoring, and so on), generally compiled as artifacts served over Uber data centers. A prediction service that receives a request and based on header information routes pre-loads the right model and feeds the prediction vector, using an internal DSL that is able to query for further data augmentation on the serving layer of the feature store.

- **Performance and monitoring**: It leverages the general centralized logging system of the company. For monitoring predictions, metrics are produced of predictions and real-world values and differences are logged. The errors of the model can in this way be analyzed and monitored.

- **Model management**: Models are compiled as artifacts and stored in a Cassandra data store.

- **Workflow management**: Provides an API for wiring the pipelines. It contains a management plane with a UI that allows the management of models and deployments. Workflow management is API-driven and can be managed in either Python or Java from the outside.

The clear advantage for a company such as Uber to have built their own system is agility and the ability to cater to their very specific use case.

Getting to know Kubeflow

Kubeflow in some way is an open source platform for the ML life cycle for **Kubernetes** environments. It's basically an ecosystem of tools that work together to provide the main components of an ML platform. Kubeflow was initially developed at Google and it's currently a very active open source project.

Kubernetes is one of the leading open source computational environments that allows flexibility in allocating computing and storage resources for containerized workloads. It was created originally at Google. In order to understand Kubeflow, a basic understanding of Kubernetes is needed. The following official documentation link contains the prerequisites to understand the basics: `https://kubernetes.io/docs/concepts/overview/`.

As shown in *Figure 6.4*, it uses the foundation of Kubernetes and provides a set of applications for the ML workflow where different tools compatible with the standards set by Kubeflow can be coalesced to provide a coherent set of services:

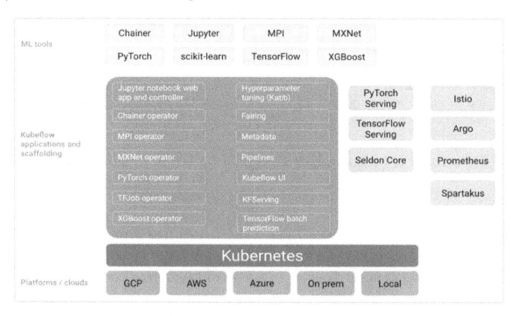

Figure 6.4 – Retrieved from https://www.kubeflow.org/docs/started/kubeflow-overview/

The components that we choose to use as a reference for architecture comparison are the following:

- **Data and feature management**: Kubeflow provides integration with big data tools such as Spark and others. A component of the ecosystem used for data and feature management is called Feast, an open source feature for ML.

- **Training infrastructure**: Kubeflow provides specific types of Kubeflow operators for common models such as, for instance, TensorFlow, PyTorch, and custom-made ones. The training jobs will basically be specific Kubernetes jobs.

- **Deployment and inference**: Kubeflow provides multiple integrations with third-party tools such as TensorFlow Serving, Seldon Core, and KFServing with different trade-offs and maturity levels.

- **Performance and monitoring**: Prometheus is a general tool used for monitoring within the Kubernetes environment and can be leveraged in this context.

- **Model management**: Not a specific tool for managing models but tools such as MLflow can be added to cover the model management life cycle.

- **Workflow management**: Workflow management is leveraged through a specific tool called Kubeflow Pipelines built on top of a generic pipeline tool for Kubernetes called Argo Workflows. It allows multi-step pipelines to be built in code.

After looking at reference architectures, we will now spend time crafting our own, armed with the state-of-the-art knowledge available in the industry.

Architecting the PsyStock ML platform

There is a set of desirable tenets that we can define for our ML platform based on a distillation of the research on best practices and example reference architectures. The main tenets that we want to maintain in our platform are the following:

- **Leverage open systems and standards**: Using open systems such as the ones available in MLflow allows longevity and flexibility to leverage the open source community advances and power to extend the company ML platform at a lower cost.

- **Favor scalable solutions**: A company needs to be prepared for a future surge in growth; although this is the first version, the ability to surge on-demand from training and perspective needs to be in place.

- **Integrated reliable data life cycle**: Data is the center of gravity of the ML platform and should be managed in a reliable and traceable manner at scale.

- **Follow SWE best practices**: For example, separation of concerns, testability, CI/CD, observability, and modularity.

- **Maintain vendor and cloud independence**: PsyStock being a start-up is operating in a very dynamic environment and in different geographies with access to different clouds and, in some cases, with compliance requirements of not moving the data from the given geography. So, being cloud-agnostic and being able to have workloads in different environments is a competitive advantage.

These tenets will allow us to frame our systems architecture within an open and low-cost solution for the company and allow the flexibility of running on the different systems on-premises, in the cloud, or local.

We have previously defined the business requirements of the prediction use cases, namely detection of the movement of cryptocurrency and value prediction. To leverage this and other use cases, the creation of an ML platform is critical to the company.

Now, armed with the knowledge from the research and description of state-of-the-art systems, we will next define eliciting the features of our ML platform.

Describing the features of the ML platform

Creating a specification of features is extremely important to keep the development efforts focused on a narrow set of features that unlock value to users of the platform. In this section, we will elicit the features that will realize the best value of an ML platform.

In our system, we want to be able to have the following features:

- **Feature: Schedule training jobs**: A data scientist needs to be able to schedule training jobs for their models using configuration or equivalent code.

- **Feature: Deploy seamlessly different models developed from the data science workbench**: The company already has a data science workbench developed in *Chapter 3, Your Data Science Workbench*. We want to be able to leverage all the work previously done so models developed on the platform can be deployed in production.

- **Feature: Allow recalibration of models in presence of new data**: When new data arrives in a specific location, a new model needs to be generated automatically and stored in a model registry accessible to systems and humans of the platform.

- **Feature: Submit and configure batch scoring jobs**: The platform should allow the relevant users to configure and schedule batch jobs in the presence of new data.

- **Feature: Efficient inference API-based scoring for the following APIs**: Given a model it should be a feature of the platform the creation of matching API using the model schema.

After discussing the ideal features of an ML system, we will start in the next section with architecting the system at a high level.

High-level systems architecture

We will now focus on defining the building blocks of our architecture and the different data flows between the different components.

Based on the features specified and tenets of the previous section, our ML platform and solution should contain the following components as described by the architecture diagram in *Figure 6.5*.

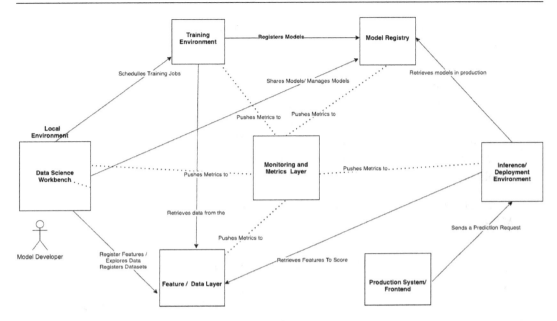

Figure.6.5 – Architectural diagram of an ML platform

This diagram is specifically agnostic from technology choices as this will be done in the respective upcoming chapters where each of the components' engineering will be explored fully.

1. **Data and feature management**: This is executed by the **Feature / Data Layer**, which receives feature registrations from the workbench and allows the registration of datasets from the workbench. The data layer provides data to the training environment.

2. **Training infrastructure**: The training infrastructure component allows the schedule of training of jobs based on a request from the data science workbench.

3. **Deployment and inference**: The deployment environment consumes models to execute either in batch or in real time prompted either by data in the data layer or by requests via production systems.

4. **Performance and monitoring**: This is accomplished through the central component of monitoring and metrics that all the systems surrounding the component publish metrics into.

5. **Model management**: Encapsulated by the component of **Model Registry**, which contains a store and the associated life cycles projects. The input comes primarily from the training jobs and the data science workbench.

6. **Workflow management**: This is a component that allows the orchestration of the different systems. For example, it allows scheduling jobs and dependency management, enforcing the order of execution. For example, an inference batch job can only be executed after a training job. This can be achieved through the operating system using a Cron system or through more sophisticated workflow tools such as Airflow.

We will next briefly touch on how we will realize the ideas outlined in this section with **MLflow**.

MLflow and other ecosystem tools

MLflow is a tool that was created by the open source community to address a gap in open systems for ML systems, focused on reproducibility, model management, and deployment. MLflow is by no means a complete tool; it needs other components and is a central part of an ML solution when its strengths are leveraged.

In recent years, systems such as **Kubeflow** have been emerging in the Kubernetes world to help manage the infrastructure side of ML systems and being the actual deployment environment.

Minio is a storage system that ships with Kubeflow that will be used as an agnostic storage mechanism for metadata and datasets and provides an abstraction for storage on both cloud and local environments.

Having identified best practices in the industry with regards to ML platforms, outlining our requirements, and, in this section, architecting our ML platform, we will spend the next four chapters of the book building each of the components of our platform.

Summary

In this chapter, we introduced the concepts involved in architecting ML systems, mapped stakeholders, identified common issues and best practices, and outlined the initial architecture. We identified critical building blocks of an ML systems architecture on the data layer and modeling and inference layer. The interconnection between the components was stressed and a specification of features was outlined.

We also addressed how MLflow can be leveraged in your ML platform and the shortcomings that can be complemented by other reference tools.

In the next chapters and section of the book, we will focus on applying the concepts learned so far to real-life systems and we will practice by implementing the architecture of the PsyStock ML platform. We will have one chapter dedicated to each component, starting from specification up to the implementation of the component with practical examples.

Further reading

In order to further your knowledge, you can consult the documentation at the following links:

- `https://www.mlflow.org/docs/latest/models.html`

- High interest of technical debt – `https://papers.nips.cc/paper/2015/file/86df7dcfd896fcaf2674f757a2463eba-Paper.pdf`

- **CS 329S**: ML systems design, *Chip Huyen* – `https://cs329s.stanford.edu`, 2021

7
Data and Feature Management

In this chapter, we will add a feature management data layer to the machine learning platform being built. We will leverage the features of the MLflow Projects module to structure our data pipeline.

Specifically, we will look at the following sections in this chapter:

- Structuring your data pipeline project
- Acquiring stock data
- Checking data quality
- Managing features

In this chapter, we will acquire relevant data to provide datasets for training. Our primary resource will be the Yahoo Finance Data for BTC dataset. Alongside that data, we will acquire the following extra datasets.

Leveraging our productionization architecture introduced in *Chapter 6, Introducing ML Systems Architecture*, represented in *Figure 7.1*, the feature and data component is responsible for acquiring data from sources and making the data available in a format consumable by the different components of the platform:

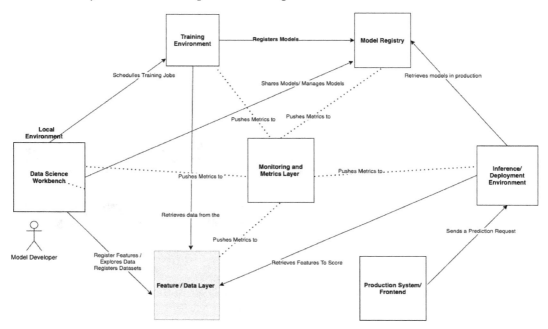

Figure 7.1 – High-level architecture with a data layer reference

Let's delve into this chapter and see how we will structure and populate the data layer with relevant data to be used for training models and generating features.

Technical requirements

For this chapter, you will need the following prerequisites:

- The latest version of Docker installed on your machine. If you don't already have it installed, please follow the instructions at `https://docs.docker.com/get-docker/`.

- The latest version of docker-compose installed. Please follow the instructions at `https://docs.docker.com/compose/install/`.

- Access to Git on the command line and installed as described at `https://git-scm.com/book/en/v2/Getting-Started-Installing-Git`.

- Access to a Bash terminal (Linux or Windows).

- Access to a browser.
- Python 3.5+ installed.
- The latest version of your machine learning installed locally as described in *Chapter 3, Your Data Science Workbench.*

In the next section, we will describe the structure of our data pipeline, the data sources, and the different steps that we will execute to implement our practical example leveraging MLflow project features to package the project.

> **Note**
>
> Copying and pasting directly from the code snippets might cause issues with your editor. Please refer to the GitHub repository of the chapter available at `https://github.com/PacktPublishing/Machine-Learning-Engineering-with-MLflow/tree/master/Chapter07`

Structuring your data pipeline project

At a high level, our data pipeline will run weekly, collecting data for the preceding 7 days and storing it in a way that can be run by machine learning jobs to generate models upstream. We will structure our data folders into three types of data:

- **Raw data**: A dataset generated by retrieving data from the Yahoo Finance API for the last 90 days. We will store the data in CSV format – the same format that it was received in from the API. We will log the run in MLflow and extract the number of rows collected.
- **Staged data**: Over the raw data, we will run quality checks, schema verification, and confirm that the data can be used in production. This information about data quality will be logged in MLflow Tracking.
- **Training data**: The training data is the final product of the data pipeline. It must be executed over data that is deemed as clean and suitable to execute models. The data contains the data processed into features that can be consumed directly for the training process.

This folder structure will be implemented initially on the filesystem and will be transposed to the relevant environment (examples: AWS S3, Kubernetes PersistentVolume, and so on) during deployment.

In order to execute our data pipeline project, we will use the **MLflow Project** module to package the data pipeline in an execution environment-independent format. We will use the Docker format to package the **MLflow Project**. The Docker format provides us with different options to deploy our project in the cloud or on-premises depending on the available infrastructure to deploy our project:

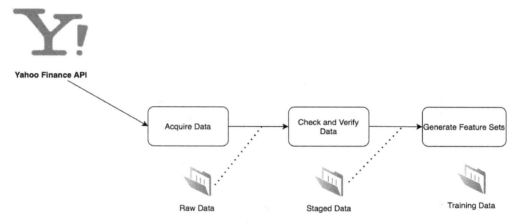

Figure 7.2 – High-level architecture with a data layer reference

Our workflow will execute the following steps, as illustrated in *Figure 7.2*:

1. **Acquire Data**: This is responsible for retrieving the data from the API and saving the data as a CSV file in the `data/raw/data.csv folder`.

2. **Check and Verify Data**: This is responsible for checking whether the data provided meets the quality requirements of the data pipeline, and if it does, it will report different metrics and write data in the `data/staged/data.csv file`.

3. **Generate Feature Sets**: Based on the staged data, this will transform the data into a format consumable by the machine learning code and produce a new training dataset at the `data/training/data.csv location`.

With these three distinct phases, we ensure the reproducibility of the training data generation process, visibility, and a clear separation of the different steps of the process.

We will start by organizing our MLflow project into steps and creating placeholders for each of the components of the pipeline:

1. Create a new folder on your local machine with the name `psytock-data-features`.

2. Add the `MLProject` file:

```
name: pystock_data_features

conda:
  file: conda.yaml

entry_points:

  data_acquisition:
    command: "python data_acquisition.py"

  clean_validate_data:
    command: "python clean_validate_data.py "

  feature_set_generation:
    command: "python feature_set_generation.py"

  main:
    command: "python main.py"
```

3. Add the following `conda.yaml` file:

```
    name: pystock-data-features
channels:
  - defaults
dependencies:
  - python=3.8
  - numpy
  - scipy
  - pandas
  - cloudpickle
  - pip:
    - git+git://github.com/mlflow/mlflow
    - pandas_datareader
    - great-expectations==0.13.15
```

4. You can now add a sample `main.py` file to the folder to ensure that the basic structure of the project is working:

```python
import mlflow
import click


def _run(entrypoint, parameters={}, source_version=None,
use_cache=True):
    #existing_run = _already_ran(entrypoint, parameters,
source_version)
    #if use_cache and existing_run:
    #    print("Found existing run for entrypoint=%s and
parameters=%s" % (entrypoint, parameters))
    #    return existing_run
    print("Launching new run for entrypoint=%s and
parameters=%s" % (entrypoint, parameters))
    submitted_run = mlflow.run(".", entrypoint,
parameters=parameters)
    return submitted_run


@click.command()
def workflow():
    with mlflow.start_run(run_name ="pystock-data-
pipeline") as active_run:
        mlflow.set_tag("mlflow.runName", "pystock-data-
pipeline")
        _run("load_raw_data")
        _run("clean_validate_data")
        _run("feature_set_generation")


if __name__=="__main__":
    workflow()
```

5. Test the basic structure by running the following command:

```
mlflow run .
```

This command will build your project based on the environment created by your `conda.yaml` file and run the basic project that you just created. It should error out as we need to add the missing files. The *file not found* error will look like the following :

```
python: can't open file 'check_verify_data.py': [Errno 2]
No such file or directory
```

At this stage, we have the basic blocks of the MLflow project of the data pipeline that we will be building in this chapter. We will next fill in the Python script to acquire the data in the next section.

Acquiring stock data

Our script to acquire the data will be based on the `pandas-datareader` Python `package`. It provides a simple abstraction to remote financial APIs we can leverage in the future in the pipeline. The abstraction is very simple. Given a data source such as Yahoo Finance, you provide the stock ticker/pair and date range, and the data is provided in a DataFrame.

We will now create the `load_raw_data.py` `file`, which will be responsible for loading the data and saving it in the `raw` folder. You can look at the contents of the file in the repository at `https://github.com/PacktPublishing/Machine-Learning-Engineering-with-MLflow/blob/master/Chapter07/psystock-data-features-main/load_raw_data.py`. Execute the following steps to implement the file:

1. We will start by importing the relevant packages:

    ```
    import mlflow
    from datetime import date
    from dateutil.relativedelta import relativedelta
    import pprint
    import pandas
    import pandas_datareader.data as web
    ```

2. Next, you should add a function to retrieve the data:

    ```
    if __name__ == "__main__":

        with mlflow.start_run(run_name="load_raw_data") as
    ```

```
run:

        mlflow.set_tag("mlflow.runName", "load_raw_data")
        end = date.today()
        start = end + relativedelta(months=-3)

        df = web.DataReader("BTC-USD", 'yahoo', start,
    end)

        df.to_csv("./data/raw/data.csv")
```

Now that we've acquired the data, we need to apply the best practices we will address in the next section – an approach to check the data quality of the data acquired.

Checking data quality

Checking data quality as part of your machine learning system is extremely critical to ensure the integrity and correctness of your model training and inference. Principles of software testing and quality should be borrowed and used on the data layer of machine learning platforms.

From a data quality perspective, in a dataset there are a couple of critical dimensions with which to assess and profile our data, namely:

- **Schema compliance**: Ensuring the data is from the expected types; making sure that numeric values don't contain any other types of data
- **Valid data**: Assessing from a data perspective whether the data is valid from a business perspective
- **Missing data**: Assessing whether all the data needed to run analytics and algorithms is available

For data validation, we will use the *Great Expectations* Python package (available at https://github.com/great-expectations/great_expectations). It allows making assertions on data with many data-compatible packages, such as pandas, Spark, and cloud environments. It provides a DSL in JSON with which to declare the rules that we want our data to be compliant with.

For our current project, we want the following rules/constraints to be verifiable:

- Date values should be valid dates and cannot be missing.

- Check numeric and long values are correctly typed.
- All columns are present in the dataset.

We will now create the `check_verify_data.py file`, which will be responsible for loading the data and saving it in the `staging` folder where all the data is valid and ready to be used for ML training. You can look at the contents of the file in the repository at `https://github.com/PacktPublishing/Machine-Learning-Engineering-with-MLflow/blob/master/Chapter07/psystock-data-features-main/check_verify_data.py`.

1. In order to convert the preceding rules so they can be relied on by our system, we will need to import the following dependencies:

```
import mlflow
from datetime import date
from dateutil.relativedelta import relativedelta
import pprint
import pandas_datareader
import pandas
from pandas_profiling import ProfileReport
import great_expectations as ge
from great_expectations.profile.basic_dataset_profiler import BasicDatasetProfiler
```

2. Next, we will implement the script:

```
if __name__ == "__main__":
    with mlflow.start_run(run_name="check_verify_data") as run:
        mlflow.set_tag("mlflow.runName", "check_verify_data")
        df = pandas.read_csv("./data/raw/data.csv")
        describe_to_dict=df.describe().to_dict()
        mlflow.log_dict(describe_to_dict,"describe_data.json")
        pd_df_ge = ge.from_pandas(df)
        assert pd_df_ge.expect_column_values_to_match_strftime_format("Date", "%Y-%m-%d").success == True
        assert pd_df_ge.expect_column_values_to_be_of_
```

```
type("High", "float").success == True
        assert pd_df_ge.expect_column_values_to_be_of_
type("Low", "float").success == True
        assert pd_df_ge.expect_column_values_to_be_of_
type("Open", "float").success == True
        assert pd_df_ge.expect_column_values_to_be_of_
type("Close", "float").success == True
        assert pd_df_ge.expect_column_values_to_be_of_
type("Volume", "long").success == True
        assert pd_df_ge.expect_column_values_to_be_of_
type("Adj Close", "float").success == True
```

3. Now we can progress to do a little bit of cleaning:

```
        #we can do some basic cleaning by dropping the
null values
        df.dropna(inplace=True)


        #if data_passes_quality_can_go_to_features:
        df.to_csv("data/staging/data.csv")
```

Having verified the quality of the data and staging to be used, it can now be utilized for feature generation with a high degree of confidence.

Generating a feature set and training data

We will refactor a bit of the code previously developed in our local environment to generate features for training to add to our MLflow project the data pipelineof our MLflow project .

We will now create the `feature_set_generation.py` file, which will be responsible for generating our features and saving them in the `training` folder where all the data is valid and ready to be used for ML training. You can look at the contents in the file in the repository `https://github.com/PacktPublishing/Machine-Learning-Engineering-with-MLflow/blob/master/Chapter07/psystock-data-features-main/feature_set_generation.py`:

1. We need to import the following dependencies:

```
import mlflow
from datetime import date
```

```
from dateutil.relativedelta import relativedelta
import pprint
import pandas as pd
import pandas_datareader
import pandas_datareader.data as web
import numpy as np
```

2. Before delving into the main component of the code, we'll now proceed to
    implement a critical function to generate the features by basically transforming the
    difference with each *n* preceding day in a feature that we will use to predict the next
    day, very similar to the approach that we used in previous chapters of the book for
    our running use case:

```
def rolling_window(a, window):
    """
        Takes np.array 'a' and size 'window' as
parameters
        Outputs an np.array with all the ordered
sequences of values of 'a' of size 'window'
        e.g. Input: ( np.array([1, 2, 3, 4, 5, 6]), 4 )
        Output:
                array([[1, 2, 3, 4],
                       [2, 3, 4, 5],
                       [3, 4, 5, 6]])
    """
    shape = a.shape[:-1] + (a.shape[-1] - window + 1,
window)
    strides = a.strides + (a.strides[-1],)
    return np.lib.stride_tricks.as_strided(a,
shape=shape, strides=strides)
```

3. Next, we'll proceed to read the staged file that is deemed as clean and ready to be
    used by upstream processes:

```
with mlflow.start_run() as run:
    mlflow.set_tag("mlflow.runName", "feature_set_
```

```
generation")
        btc_df = pd.read_csv("data/staging/data.csv")
        btc_df['delta_pct'] = (btc_df['Close'] - btc_
df['Open'])/btc_df['Open']
        btc_df['going_up'] = btc_df['delta_pct'].
apply(lambda d: 1 if d>0.00001 else 0).to_numpy()
        element=btc_df['going_up'].to_numpy()
        WINDOW_SIZE=15
        training_data = rolling_window(element, WINDOW_
SIZE)
        pd.DataFrame(training_data).to_csv("data/
training/data.csv", index=False)
```

We generate the feature set and features. We are now able to run all of the end-to-end pipeline from data acquisition to feature generation.

Running your end-to-end pipeline

In this section, we will run the complete example, which you can retrieve from the following address for the book's GitHub repository in the folder at /Chapter07/ psytock-data-features-main. *Figure 7.3* illustrates the complete folder structure of the project that you can inspect in GitHub and compare with your local version:

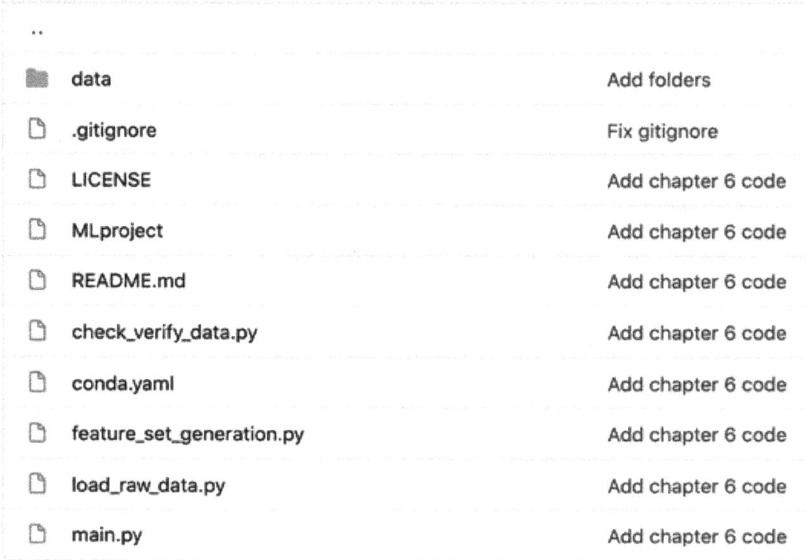

..		
📁	data	Add folders
🗋	.gitignore	Fix gitignore
🗋	LICENSE	Add chapter 6 code
🗋	MLproject	Add chapter 6 code
🗋	README.md	Add chapter 6 code
🗋	check_verify_data.py	Add chapter 6 code
🗋	conda.yaml	Add chapter 6 code
🗋	feature_set_generation.py	Add chapter 6 code
🗋	load_raw_data.py	Add chapter 6 code
🗋	main.py	Add chapter 6 code

Figure 7.3 – Folder structure

To run the pipeline end to end, you should execute the following command in the directory with the code:

```
mlflow run . --experiment-name=psystock_data_pipelines
```

It will basically execute the end-to-end pipeline and you can inspect it directly in the MLflow UI, running each of the steps of the pipeline in order:

```
mlflow ui
```

You can run and explore the tracking information in MLflow at `http://localhost:5000`.

In *Figure 7.4*, you can see the different runs of the main project and subprojects of the stages of the pipeline in a nested workflow format that you can browse to inspect the details:

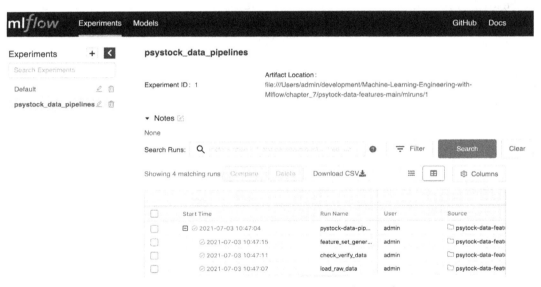

Figure 7.4 – High-level architecture with a data layer reference

In *Figure 7.5*, you can see the reference to the load_raw_data phase of the data pipeline and check when it was started and stopped and the parameters used:

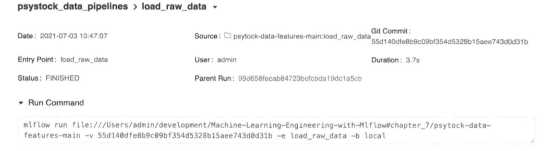

Figure 7.5 – High-level architecture with a data layer reference

In *Figure.7.6*, you can see the reference to the check_verify_data phase of the data pipeline where we logged some basic statistical information of the dataset obtained:

Figure 7.6 – High-level architecture with a data layer reference

If any data quality issues are detected, the workflow will fail with a clear indication of which section failed, as represented in *Figure 7.7*:

	Start Time	Run Name	Source	Version
☐	⊟ ⊗ 2021-07-03 11:12:50	pystock-data-pipeline	⬚ psytock-data-features-	55d140
☐	⊗ 2021-07-03 11:13:01	feature_set_generation	⬚ psytock-data-features-	55d140
☐	⊘ 2021-07-03 11:12:56	check_verify_data	⬚ psytock-data-features-	55d140
☐	⊘ 2021-07-03 11:12:53	load_raw_data	⬚ psytock-data-features-	55d140
	Load more			

Figure 7.7 – Checking errors

With this section, we have concluded the description of the process of data management and feature generation in a data pipeline implemented with the `MLProjects` module in MLflow. We will now look at how to manage the data in a feature store.

Using a feature store

A feature store is a software layer on top of your data to abstract all the production and management processes for data by providing inference systems with an interface to retrieve a feature set that can be used for inference or training.

In this section, we will illustrate the concept of a feature store by using Feast (a feature store), an operational data system for managing and serving machine learning features to models in production:

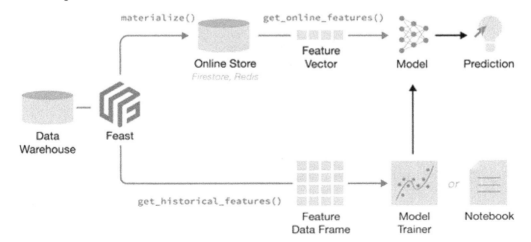

Figure 7.8 – Feast Architecture (retrieved from https://docs.feast.dev/)

In order to understand how Feast works and how it can fit into your data layer component (code available at `https://github.com/PacktPublishing/Machine-Learning-Engineering-with-MLflow/tree/master/Chapter07/psystock_feature_store`, execute the following steps:

1. Install `feast`:

    ```
    pip install feast==0.10
    ```

2. Initialize a feature repository:

    ```
    feast init
    ```

3. Create your feature definitions by replacing the `yaml` file generated automatically:

    ```
    project: psystock_feature_store
    registry: data/registry.db
    provider: local
    online_store:
        path: data/online_store.db
    ```

4. We will now proceed to import dependencies of the feature definition:

    ```
    from google.protobuf.duration_pb2 import Duration
    from feast import Entity, Feature, FeatureView, ValueType
    from feast.data_source import FileSource
    ```

5. We can now load the feature files:

    ```
    token_features = FileSource(
        path="/data/features.csv",
        event_timestamp_column="create_date",
        created_timestamp_column="event_date",
    )


    token= Entity(name="token", value_type=ValueType.STRING,
    description="token id",)
    ```

6. We can now add a feature view:

    ```
    hourly_view_features_token = FeatureView(
        name="token_hourly_features",
        entities=["token"],
    ```

```
        ttl=Duration(seconds=3600 * 1),
        features=[
            Feature(name="prev_10days", dtype=ValueType.
    INT64),
            Feature(name="prev_11days", dtype=ValueType.
    INT64),
            Feature(name="prev_12days", dtype=ValueType.
    INT64),
            Feature(name="prev_13days", dtype=ValueType.
    INT64)
        ],
        online=True,
        input=token_features,
        tags={},
    )
```

7. To deploy the feature store with the configurations added so far, we need to run the following command:

```
feast apply
```

At this stage, the feature store is deployed in your environment (locally in this case) and the feature store is available to be used from your MLflow job.

8. We can now do feature retrieval, now that all the features are stored in a feature store:

```
import pandas as pd
from datetime import datetime
from feast import FeatureStore


# entity_df generally comes from upstream systems
event_data_point = pd.DataFrame.from_dict({
    "token": ["btc","btc"],
    "event_date": [
        datetime(2021, 4, 12, 10, 59, 42),
        datetime(2021, 4, 12, 8,  12, 10),
    ]
```

```
    })
    store = FeatureStore(repo_path=".")

    feature_loading_df = store.get_historical_features(
        entity_df=entity_df,
        feature_refs = [
            'token_hourly_features:prev_3days',
            'token_hourly_features:prev_4days',
            'token_hourly_features:prev_5days'
        ],
    ).to_df()
```

You can now integrate your feature store repository into your MLflow workloads.

With this section, we have concluded the description of the process of data management and feature generation in a data pipeline implemented with the MLProjects module in MLflow. We are now ready to deal with production environment deployments in subsequent chapters.

Summary

In this chapter, we covered MLflow and its integration with the feature management data layer of our reference architecture. We leveraged the features of the MLflow Projects module to structure our data pipeline.

The important layer of data and feature management was introduced, and the need for feature generation was made clear, as were the concepts of data quality, validation, and data preparation.

We applied the different stages of producing a data pipeline to our own project. We then formalized data acquisition and quality checks. In the last section, we introduced the concept of a feature store and how to create and use one.

In the next chapters and following section of the book, we will focus on applying the data pipeline and features to the process of training and deploying the data pipeline in production.

Further reading

In order to further your knowledge, you can consult the documentation at the following link:

`https://github.com/mlflow/mlflow/blob/master/examples/multistep_workflow/MLproject`

8
Training Models with MLflow

In this chapter, you will learn about creating production-ready training jobs with MLflow. In the bigger scope of things, we will focus on how to move from the training jobs in the notebook environment that we looked at in the early chapters to a standardized format and blueprint to create training jobs.

Specifically, we will look at the following sections in this chapter:

- Creating your training project with MLflow
- Implementing the training job
- Evaluating the model
- Deploying the model in the Model Registry
- Creating a Docker image for your training job

It's time to add to the pyStock **machine learning** (**ML**) platform training infrastructure to take **proof-of-concept** models created in the workbench developed in *Chapter 3, Your Data Science Workbench to a Production Environment*.

In this chapter, you will be developing a training project that runs periodically or when triggered by a dataset arrival. The main output of the training project is a new model that is generated as output and registered in the Model Registry with different details.

Here is an overview of the training workflow:

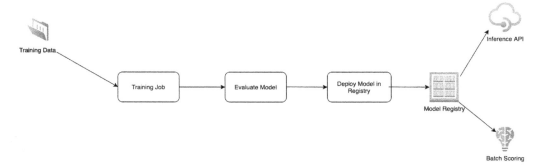

Figure 8.1 – Training workflow

Figure 8.1 describes at a high level the general process, whereby a training dataset arrives and a training job kicks in. The training job produces a model that is finally evaluated and deployed in the Model Registry. Systems upstream are now able to deploy inference **application programming interfaces (APIs)** or batch jobs with the newly deployed model.

Technical requirements

For this chapter, you will need the following prerequisites:

- The latest version of Docker installed on your machine. If you don't already have it installed, please follow the instructions at `https://docs.docker.com/get-docker/`.

- The latest version of Docker Compose installed—please follow the instructions at `https://docs.docker.com/compose/install/`.

- Access to Git in the command line, and installed as described at `https://git-scm.com/book/en/v2/Getting-Started-Installing-Git`.

- Access to a Bash terminal (Linux or Windows).

- Access to a browser.

- Python 3.5+ installed.

- The latest version of your ML library installed locally as described in *Chapter 4, Experiment Management in MLflow*

Creating your training project with MLflow

You receive a specification from a data scientist based on the **XGBoost** model being ready to move from a **proof-of-concept** to a production phase.

We can review the original Jupyter notebook from which the model was registered initially by the data scientist, which is a starting point to start creating an ML engineering pipeline. After initial prototyping and training in the notebook, they are ready to move to production.

Some companies go directly to productionize the notebooks themselves and this is definitely a possibility, but it becomes impossible for the following reasons:

- It's hard to version notebooks.
- It's hard to unit-test the code.
- It's unreliable for long-running tests.

With these three distinct phases, we ensure reproducibility of the training data-generation process and visibility and clear separation of the different steps of the process.

We will start by organizing our MLflow project into steps and creating placeholders for each of the components of the pipeline, as follows:

1. Start a new folder in your local machine and name this `pystock-training`. Add the `MLProject` file, as follows:

```
name: pystock_training

conda_env: conda.yaml

entry_points:

  main:
    data_file: path
    command: "python main.py"

  train_model:
    command: "python train_model.py"

  evaluate_model:
    command: "python evaluate_model.py "
```

```
register_model:
    command: "python register_model.py"
```

2. Add the following `conda.yaml` file:

```
name: pystock-training
channels:
  - defaults
dependencies:
  - python=3.8
  - numpy
  - scipy
  - pandas
  - cloudpickle
  - pip:
    - git+git://github.com/mlflow/mlflow
    - sklearn
    - pandas_datareader
    - great-expectations==0.13.15
    - pandas-profiling
    - xgboost
```

3. You can add now a sample `main.py` file to the folder to ensure that the basic structure of the project is working, as follows:

```
import mlflow
import click
import os


def _run(entrypoint, parameters={}, source_version=None,
use_cache=True):
    print("Launching new run for entrypoint=%s and
parameters=%s" % (entrypoint, parameters))
    submitted_run = mlflow.run(".", entrypoint,
parameters=parameters)
```

```
        return mlflow.tracking.MlflowClient().get_
run(submitted_run.run_id)


@click.command()
def workflow():
    with mlflow.start_run(run_name ="pystock-training")
as active_run:
        mlflow.set_tag("mlflow.runName", "pystock-
training")
        _run("train_model")
        _run("evaluate_model")
        _run("register_model")

if __name__=="__main__":
    workflow()
```

4. Test the basic structure by running the following command:

    ```
    mlflow run.
    ```

    This command will build your project based on the environment created by your `conda.yaml` file and run the basic project you just created. It should error out, as we need to add the missing files.

At this stage, we have the basic blocks of the MLflow project of the data pipeline that we will be building in this chapter. You will next fill in the Python file to train the data.

Implementing the training job

We will use the training data produced in the previous chapter. The assumption here is that an independent job populates the data pipeline in a specific folder. In the book's GitHub repository, you can look at the data in `https://github.com/PacktPublishing/Machine-Learning-Engineering-with-MLflow/blob/master/Chapter08/psystock-training/data/training/data.csv`.

We will now create a `train_model.py` file that will be responsible for loading the training data to fit and produce a model. Test predictions will be produced and persisted in the environment so that other steps of the workflow can use the data to evaluate the model.

The file produced in this section is available at the following link:

https://github.com/PacktPublishing/Machine-Learning-Engineering-with-MLflow/blob/master/Chapter08/psystock-training/train_model.py:

1. We will start by importing the relevant packages. In this case, we will need `pandas` to handle the data, `xgboost` to run the training algorithm, and—obviously—`mlflow` to track and log the data run. Here is the code you'll need to do this:

```
import pandas as pd
import mlflow
import xgboost as xgb
import mlflow.xgboost
from sklearn.model_selection import train_test_split
```

2. Next, you should add a function to execute the split of the data relying on `train_test_split` from `sklearn`. Our chosen split is 33/67% for testing and training data respectively. We specify the `random_state` parameter in order to make the process reproducible, as follows:

```
def train_test_split_pandas(pandas_df,t_size=0.33,r_state=42):
    X=pandas_df.iloc[:,:-1]
    Y=pandas_df.iloc[:,-1]
    X_train, X_test, y_train, y_test = train_test_split(X, Y, test_size=t_size, random_state=r_state)

    return X_train, X_test, y_train, y_test
```

3. This function returns the train and test dataset and the targets for each dataset. We rely on the `xgboost` matrix `xgb.Dmatrix` data format to efficiently load the training and testing data and feed the `xgboost.train` method. The code is illustrated in the following snippet:

```
if __name__ == "__main__":

    THRESHOLD = 0.5

    mlflow.xgboost.autolog()
```

```
with mlflow.start_run(run_name="train_model") as run:
    mlflow.set_tag("mlflow.runName", "train_model")


    pandas_df=pd.read_csv("data/training/data.csv")

    pandas_df.reset_index(inplace=True)

    X_train, X_test, y_train, y_test = train_test_
split_pandas(pandas_df)

    train_data = xgb.DMatrix(X_train, label=y_train)
    test_data = xgb.DMatrix(X_test)

    model = xgb.train(dtrain=train_data,params={})
```

4. We also use this moment to produce test predictions using the `model.predict` method. Some data transformation is executed to discretize the probability of the stock going up or down and transform it into 0 (not going up) or 1 (going up), as follows:

```
y_probas=model.predict(test_data)
y_preds = [1 if y_proba > THRESHOLD else 0. for
y_proba in y_probas]
```

5. As a last step, we will persist the test predictions on the `result` variable. We drop the index so that the saved `pandas` DataFrame doesn't include the index when running the `result.to_csv` command, as follows:

```
test_prediction_results = pd.DataFrame(data={'y_
pred':y_preds,'y_test':y_test})

result = test_prediction_results.reset_
index(drop=True)

result.to_csv("data/predictions/test_predictions.
csv")
```

6. You can look at your MLflow **user interface** (**UI**) by running the following command to see the metrics logged:

```
mlflow ui
```

You should be able to look at your MLflow UI, available to view in the following screenshot, where you can see the persisted model and the different model information of the just-trained model:

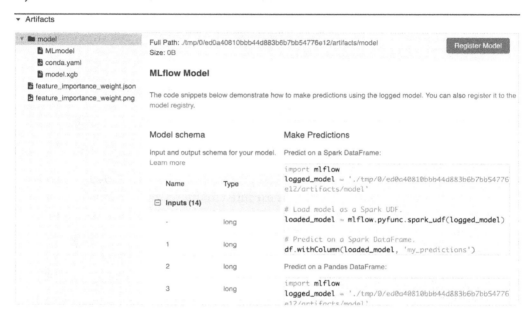

Figure 8.2 – Training model

At this stage, we have our model saved and persisted on the artifacts of our MLflow installation. We will next add a new step to our workflow to produce the metrics of the model just produced.

Evaluating the model

We will now move on to collect evaluation metrics for our model, to add to the metadata of the model.

We will work on the `evaluate_model.py` file. You can follow along by working in an empty file or by going to `https://github.com/PacktPublishing/Machine-Learning-Engineering-with-MLflow/blob/master/Chapter08/psystock-training/evaluate_model.py`. Proceed as follows:

1. Import the relevant packages—`pandas` and `mlflow`—for reading and running the steps, respectively. We will rely on importing a selection of model-evaluation metrics available in `sklearn` for classification algorithms, as follows:

```
import pandas as pd
import mlflow
from sklearn.model_selection import train_test_split
from sklearn.metrics import \
    classification_report, \
    confusion_matrix, \
    accuracy_score, \
    auc, \
    average_precision_score, \
    balanced_accuracy_score, \
    f1_score, \
    fbeta_score, \
    hamming_loss, \
    jaccard_score, \
    log_loss, \
    matthews_corrcoef, \
    precision_score, \
    recall_score, \
    zero_one_loss
```

At this stage, we have imported all the functions we need for the metrics we need to extract in the next section.

2. Next, you should add a `classification_metrics` function to generate metrics based on a `df` parameter. The assumption is that the DataFrame has two columns: `y_pred`, which is the target predicted by the training model, and `y_test`, which is the target present on the training data file. Here is the code you will need:

```
def classification_metrics(df:None):
    metrics={}
    metrics["accuracy_score"]=accuracy_score(df["y_
```

```
pred"], df["y_test"]  )
    metrics["average_precision_score"]=average_precision_
score( df["y_pred"], df["y_test"]  )
    metrics["f1_score"]=f1_score( df["y_pred"], df["y_
test"]  )
    metrics["jaccard_score"]=jaccard_score( df["y_pred"],
df["y_test"]  )
    metrics["log_loss"]=log_loss( df["y_pred"], df["y_
test"]  )
    metrics["matthews_corrcoef"]=matthews_corrcoef(
df["y_pred"], df["y_test"]  )
    metrics["precision_score"]=precision_score( df["y_
pred"], df["y_test"]  )
    metrics["recall_score"]=recall_score( df["y_pred"],
df["y_test"]  )
    metrics["zero_one_loss"]=zero_one_loss( df["y_pred"],
df["y_test"]  )
    return metrics
```

The preceding function produces a `metrics` dictionary based on the predicted values and the test predictions.

3. After creating this function that generates the metrics, we need to use `start_run`, whereby we basically read the prediction test file and run the metrics. We post all the metrics in **MLflow** by using the `mlflow.log_metrics` method to log a dictionary of multiple metrics at the same time. The code is illustrated in the following snippet:

```
if __name__ == "__main__":

    with mlflow.start_run(run_name="evaluate_model") as
run:
        mlflow.set_tag("mlflow.runName", "evaluate_
model")
        df=pd.read_csv("data/predictions/test_
predictions.csv")
        metrics = classification_metrics(df)
        mlflow.log_metrics(metrics)
```

4. We can look again at the MLflow UI, where we can see the different metrics just persisted. You can view the output here:

▼ Metrics

Name	Value
accuracy_score 📈	0.423
average_precision_score 📈	0.431
f1_score 📈	0.444
jaccard_score 📈	0.286
log_loss 📈	19.93
matthews_corrcoef 📈	-0.144
precision_score 📈	0.4
recall_score 📈	0.5
zero_one_loss 📈	0.577

Figure 8.3 – Training model metrics persisted

At this stage, we have a model evaluation for our training job, providing metrics and information to model implementers/deployers. We will now move on to the last step of the training process, which is to register the model in the MLflow Model Registry so that it can be deployed in production.

Deploying the model in the Model Registry

Next, you should add the `register_model.py` function to register the model in the Model Registry.

This is as simple as executing the `mlflow.register_model` method with the **Uniform Resource Identifier (URI)** of the model and the name of the model. Basically, a model will be created if it doesn't already exist. If it's already in the registry, a new version will be added, allowing the deployment tools to look at the models and trace the training jobs and metrics. It also allows a decision to be made as to whether to promote the model to production or not. The code you'll need is illustrated in the following snippet:

```python
import mlflow

if __name__ == "__main__":

    with mlflow.start_run(run_name="register_model") as run:

        mlflow.set_tag("mlflow.runName", "register_model")

        model_uri = "runs:/{}/sklearn-model".format(run.info.run_id)

        result = mlflow.register_model(model_uri, "training-model-psystock")
```

In the following screenshot, the registered model is presented, and we can change state and move into staging or production, depending on our workflow:

Figure 8.4 – Registered model

After having registered our model, we will now move on to prepare a Docker image of our training job that can be used in a public cloud environment or in a Kubernetes cluster.

Creating a Docker image for your training job

A Docker image is, in many contexts, the most critical deliverable of a model developer to a more specialized systems infrastructure team in production for a training job. The project is contained in the following folder of the repository: https://github.com/PacktPublishing/Machine-Learning-Engineering-with-MLflow/tree/master/Chapter08/psystock-training-docker. In the following steps, we will produce a ready-to-deploy Docker image of the code produced:

1. You need to set up a Docker file in the root folder of the project, as shown in the following code snippet:

```
FROM continuumio/miniconda3:4.9.2

RUN apt-get update && apt-get install build-essential -y

RUN pip install \
    mlflow==1.18.0 \
    pymysql==1.0.2 \
    boto3

COPY ./training_project /src

WORKDIR /src
```

2. We will start by building and training the image by running the following command:

```
docker build -t psystock_docker_training_image .
```

3. You can run your image, specifying your tracking server **Uniform Resource Locator (URL)**. If you are using a local address for your MLflow Tracking Server to test the newly created image, you can use the `$TRACKING_SERVER_URI` value to reach `http://host.docker.internal:5000`, as illustrated in the following code snippet:

```
docker run -e MLflow_TRACKING_SERVER=$TRACKING_SERVER_URI
psystock_docker_training_image
```

At this stage, we have concluded all the steps of our complete training workflow. In the next chapter, we will proceed to deploy the different components of the platform in production environments, leveraging all the MLflow projects created so far.

Summary

In this chapter, we introduced the concepts and different features in terms of using MLflow to create production training processes.

We started by setting up the basic blocks of the MLflow training project and followed along throughout the chapter to, in sequence, train a model, evaluate a trained model, and register a trained model. We also delved into the creation of a ready-to-use image for your training job.

This was an important component of the architecture, and it will allow us to build an end-to-end production system for our ML system in production. In the next chapter, we will deploy different components and illustrate the deployment process of models.

Further reading

In order to further your knowledge, you can consult the official documentation at the following link:

`https://www.mlflow.org/docs/latest/projects.html`

9
Deployment and Inference with MLflow

In this chapter, you will learn about an end-to-end deployment infrastructure for our **Machine Learning (ML)** system including the inference component with the use of MLflow. We will then move to deploy our model in a cloud-native ML system (AWS SageMaker) and in a hybrid environment with Kubernetes. The main goal of the exposure to these different environments is to equip you with the skills to deploy an ML model under the varying environmental (cloud-native, and on-premises) constraints of different projects.

The core of this chapter is to deploy the PsyStock model to predict the price of Bitcoin (BTC/USD) based on the previous 14 days of market behavior that you have been working on so far throughout the book. We will deploy this in multiple environments with the aid of a workflow.

Specifically, we will look at the following sections in this chapter:

- Starting up a local model registry
- Setting up a batch inference job

- Creating an API process for inference
- Deploying your models for batch scoring in Kubernetes
- Making a cloud deployment with AWS SageMaker

Technical requirements

For this chapter, you will need the following prerequisites:

- The latest version of Docker installed on your machine. If you don't already have it installed, please follow the instructions at `https://docs.docker.com/get-docker/`.
- The latest version of `docker-compose` installed. Please follow the instructions at `https://docs.docker.com/compose/install/`.
- Access to Git in the command line, which can be installed as described at `https://git-scm.com/book/en/v2/Getting-Started-Installing-Git`.
- Access to a Bash terminal (Linux or Windows).
- Access to a browser.
- Python 3.5+ installed.
- The latest version of your ML platform installed locally as described in *Chapter 3, Your Data Science Workbench*.
- An AWS account configured to run the MLflow model.

Starting up a local model registry

Before executing the following sections in this chapter, you will need to set up a centralized model registry and tracking server. We don't need the whole of the Data Science Workbench, so we can go directly to a lighter variant of the workbench built into the model that we will deploy in the following sections. You should be in the root folder of the code for this chapter, available at `https://github.com/PacktPublishing/Machine-Learning-Engineering-with-MLflow/tree/master/Chapter09`

Next, move to the `gradflow` directory and start a light version of the environment to serve your model, as follows:

```
$ cd gradflow
$ export MLFLOW_TRACKING_URI=http://localhost:5000
$ make gradflow-light
```

After having set up our infrastructure for API deployment with MLflow with the model retrieved from the ML registry, we will next move on to the cases where we need to score some batch input data. We will prepare a batch inference job with MLflow for the prediction problem at hand.

Setting up a batch inference job

The code required for this section is in the `pystock-inference-api folder`. The MLflow infrastructure is provided in the Docker image accompanying the code as shown in the following figure:

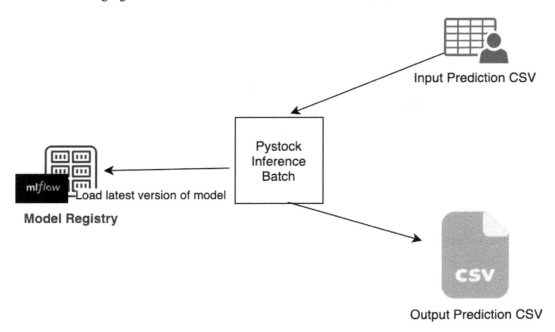

Figure 9.1 – Layout of a batch scoring deployment

If you have direct access to the artifacts, you can do the following. The code is available under the `pystock-inference-batch` directory. In order to set up a batch inference job, we will follow these steps:

1. Import the dependencies of your batch job; among the relevant dependencies we include `pandas`, `mlflow`, and `xgboost`:

```
import pandas as pd
import mlflow
import xgboost as xgb
import mlflow.xgboost
import mlflow.pyfunc
```

2. We will next load `start_run` by calling `mlflow.start_run` and load the data from the `input.csv` scoring input file:

```
if __name__ == "__main__":
    with mlflow.start_run(run_name="batch_scoring") as
run:


        data=pd.read_csv("data/input.csv",header=None)
```

3. Next, we load the model from the registry by specifying the `model_uri` value, based on the details of the model:

```
model_name = "training-model-psystock"
stage = 'Production'
model = mlflow.pyfunc.load_model(
        model_uri=f"models:/{model_name}/{stage}"
)
```

4. We are now ready to predict over the dataset that we just read by running `model.predict`:

```
y_probas=model.predict(data)
```

5. Save the batch predictions. This basically involves mapping the probability target (of the market going up) in the `y_preds` variable to a value ranging from 0 to 1:

```
y_preds = [1 if y_proba > 0.5 else 0 for y_proba in
y_probas]
```

```
data[len(data.columns)] =y_preds

result = data

result.to_csv("data/output.csv")
```

6. We now need to package the job as a Docker image so we can run it in production easily:

```
FROM continuumio/miniconda3

WORKDIR /batch-scoring/

RUN pip install mlflow==1.16.0

RUN pip install pandas==1.2.4

COPY batch_scoring.py    /batch-scoring/
COPY MLproject           /batch-scoring/

ENV MLFLOW_TRACKING_URI=http://localhost:5000

ENTRYPOINT ["mlflow run . --no-conda"]
```

7. Build your Docker image and tag it so you can reference it:

```
docker build . -t pystock-inference-batch
```

8. Run your Docker image by executing the following command:

```
docker run -i pystock-inference-batch
```

A Docker image in this case provides you with a mechanism to run your batch scoring job in any computing environment that supports Docker images in the cloud or on-premises.

We will move now to illustrate the generation of a dockerized API inference environment for MLflow.

Creating an API process for inference

The code required for this section is in the `pystock-inference-api folder`. The MLflow infrastructure is provided in the Docker image accompanying the code as shown in the following figure:

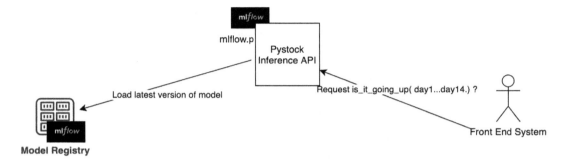

Figure 9.2 – The structure of the API job

Setting up an API system is quite easy by relying on the MLflow built-in REST API environment. We will rely on the artifact store on the local filesystem to test the APIs.

With the following set of commands, which at its core consists of using the `models` `serve` command in the CLI, we can serve our models:

```
cd /gradflow/
export MLFLOW_TRACKING_URI=http://localhost:5000
mlflow models serve -m "models:/training-model-psystock/
Production" -p 6000
```

We next will package the preceding commands in a Docker image so it can be used on any environment for deployment. The steps to achieve this are the following:

1. Generate a Docker image specifying the work directory and the commands that need to be started as an `entry point`:

```
FROM continuumio/miniconda3


WORKDIR /batch-scoring/
RUN pip install mlflow==1.16.0


ENV MLFLOW_TRACKING_URI=http://localhost:5000
```

```
ENTRYPOINT ["mlflow models serve -m "models:/training-
model-psystock/Production" -p 6000"]
```

2. Build your Docker image:

```
docker build . -t pystock-inference-api
```

3. Run your Docker image:

```
docker run -i pystock-inference-api -p 6000:6000
```

At this stage, you have dockerized the API infrastructure and can deploy it on a compute environment convenient to you.

After having delved into the interaction of MLflow and a cloud-native deployment on the AWS platform, we will now look at a deployment that is independent of any provider.

Deploying your models for batch scoring in Kubernetes

We will use Kubernetes to deploy our batch scoring job. We will need to do some modifications to make it conform to the Docker format acceptable to the MLflow deployment in production through Kubernetes. The prerequisite of this section is that you have access to a Kubernetes cluster or can set up a local one. Guides for this can be found at `https://kind.sigs.k8s.io/docs/user/quick-start/` or `https://minikube.sigs.k8s.io/docs/start/`.

You will now execute the following steps to deploy your model from the registry in Kubernetes:

1. Prerequisite: Deploy and configure `kubectl` (`https://kubernetes.io/docs/reference/kubectl/overview/`) and link it to your Kubernetes cluster.

2. Create a Kubernetes backend configuration file:

```
{
    "kube-context": "docker-for-desktop",
    "repository-uri": "username/mlflow-kubernetes-example",
    "kube-job-template-path": "/Users/username/path/to/
kubernetes_job_template.yaml"
}
```

3. Load the input files and run the model:

```
mlflow run . --backend kubernetes --backend-config
kubernetes_config.json
```

Having looked at deploying models in Kubernetes, we will now focus on deploying our model in a cloud-native ML platform.

Making a cloud deployment with AWS SageMaker

In the last few years, services such as AWS SageMaker have been gaining ground as an engine to run ML workloads. MLflow provides integrations and easy-to-use commands to deploy your model into the SageMaker infrastructure. The execution of this section will take several minutes (5 to 10 minutes depending on your connection) due to the need to build large Docker images and push the images to the Docker Registry.

The following is a list of some critical prerequisites for you to follow along:

- The AWS CLI configured locally with a default profile (for more details, you can look at https://docs.aws.amazon.com/cli/latest/userguide/cli-chap-configure.html).

- AWS access in the account to SageMaker and its dependencies.

- AWS access in the account to push to Amazon **Elastic Container Registry** (**ECR**) service.

- Your MLflow server needs to be running as mentioned in the first *Starting up a local model registry* section.

To deploy the model from your local registry into AWS SageMaker, execute the following steps:

1. Build your `mlflow-pyfunc` image. This is the basic image that will be compatible with SageMaker.

2. Build and push a container with an `mlflow pyfunc` message:

```
mlflow sagemaker build-and-push-container
```

This command will build the MLflow default image and deploy it to the Amazon ECR container.

In order to confirm that this command was successful, you can check your ECR instance on the console:

Figure 9.3 – SageMaker deployed image

3. Run your model locally to test the SageMaker Docker image and export the tracking URI:

```
$ export MLFLOW_TRACKING_URI=http://localhost:5000
mlflow sagemaker run-local -m models:/training-model-
psystock/Production  -p 7777
```

This will basically load your model into the generic API for model inference and will allow you to run the model locally to test whether it works before deploying it in the cloud. It will run the model under port 7777.

The output should look like the following excerpt and you should be able to test your model locally:

```
Installing collected packages: mlflow
  Attempting uninstall: mlflow
    Found existing installation: mlflow 1.16.0
    Uninstalling mlflow-1.16.0:
      Successfully uninstalled mlflow-1.16.0
Successfully installed mlflow-1.15.0
pip 20.2.4 from /miniconda/lib/python3.8/site-packages/
pip (python 3.8)
Python 3.8.5
1.15.0
```

```
[2021-05-08 14:01:43 +0000] [354] [INFO] Starting
gunicorn 20.1.0
[2021-05-08 14:01:43 +0000] [354] [INFO] Listening at:
http://127.0.0.1:8000 (354)
```

This will basically confirm that the image is working as expected and you should be able to run your API in SageMaker.

4. Double-check your image through the AWS `cli`:

```
aws ecr describe-images --repository-name mlflow-pyfunc
```

You should see your deployed image in the list of images and definitely ready to run.

5. You need to configure a role in AWS as specified that allows SageMaker to create resources on your behalf (you can find more details at `https://docs.databricks.com/administration-guide/cloud-configurations/aws/sagemaker.html#step-1-create-an-aws-iam-role-and-attach-sagemaker-permission-policy`).

6. Next, you need to export your region and roles to the $REGION and $ROLE environment variables with the following commands, specifying the actual values of your environment:

```
export $REGION=your-aws-region
export $ROLE=your sagemaker-enabled-role
```

7. To deploy your model to SageMaker, run the following command:

```
mlflow sagemaker deploy -a pystock-api -m models:/
training-model-psystock/Production -region-name $REGION
-- $ROLE
```

This command will load your model from your local registry into SageMaker as an internal representation and use the generated Docker image to serve the model in the AWS SageMaker infrastructure engine. It will take a few minutes to set up all the infrastructure. Upon success, you should see the following message:

```
2021/05/08 21:09:12 INFO mlflow.sagemaker: The deployment
operation completed successfully with message: "The
SageMaker endpoint was created successfully."
```

8. Verify your SageMaker endpoint:

```
aws sagemaker list-endpoints
```

You can look at the following for an illustrative example of the type of output message:

```
{
    "Endpoints": [
        {
            "EndpointName": "pystock-api",
            "EndpointArn": "arn:aws:sagemaker:eu-west-
1:123456789:endpoint/pystock-api",
            "CreationTime": "2021-05-
08T21:01:13.130000+02:00",
            "LastModifiedTime": "2021-05-
08T21:09:08.947000+02:00",
            "EndpointStatus": "InService"
        }
    ]
}
```

9. Next we need to consume our API with a simple script that basically the features, invokes the SageMaker endpoint using the Amazon Boto3 client, and prints the probablity of the market pricesgiven the feature vector:

```
import pandas
import boto3
features = pd.DataFrame([[1,0,1,1,0,1,0,1,0,1,0,1,0,1]])
payload = features.to_json(orient="split")
result  = runtime.invoke_endpoint(
            EndpointName='pystock-api', Body=payload,
            ContentType='application/json')
preds = result['Body'].read().decode("ascii")
print(preds)
```

After running this previous script you should see the following output:

```
'[0.04279635474085808]
```

10. Explore the SageMaker endpoint interface. In its monitoring component, you can look at different metrics related to your deployment environment and model as shown in *Figure 9.4*:

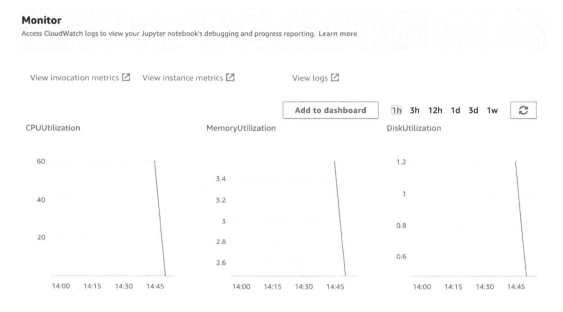

Figure 9.4 – SageMaker inference instance metrics

11. You can now easily tear down your deployed model, when in need to deploy the model or phase it out. All associated resources will be torn down:

```
mlflow sagemaker delete -a pystock-api --region-name
$REGION
```

Upon deletion, you should see a message similar to the one in the following excerpt:

```
2021/05/08 23:49:46 INFO mlflow.sagemaker: The deletion
operation completed successfully with message: "The
SageMaker endpoint was deleted successfully."
```

```
2021/05/08 23:49:46 INFO mlflow.sagemaker: Cleaning up
unused resources...
```

```
2021/05/08 23:49:47 INFO mlflow.sagemaker: Deleted
associated endpoint configuration with arn:
arn:aws:sagemaker:eu-west-1:123456789:endpoint-config/
pystock-api-config-v-hznm3ttxwx-g8uavbzia
```

```
2021/05/08 23:49:48 INFO mlflow.sagemaker: Deleted
associated model with arn: arn:aws:sagemaker:eu-
west-1:123456789:model/pystock-api-model-
4nly3634reqomejx1owtdg
```

With this section, we concluded the description of the features related to deploying an ML model with MLflow in production in different environments from your local machine, including Docker and `docker-compose`, public clouds, and the very flexible approach of using AWS SageMaker.

Summary

In this chapter, we focused on production deployments of ML models, the concepts behind this, and the different features available for deploying in multiple environments with MLflow.

We explained how to prepare Docker images ready for deployment. We also clarified how to interact with Kubernetes and AWS SageMaker to deploy models.

In the next chapter and the following sections of the book, we will focus on using tools to help scale out our MLflow workloads to improve the performance of our models' infrastructure.

Further reading

In order to further your knowledge, you can consult the documentation at the following links:

- `https://www.mlflow.org/docs/latest/python_api/mlflow.sagemaker.html`

- `https://aws.amazon.com/blogs/machine-learning/managing-your-machine-learning-lifecycle-with-mlflow-and-amazon-sagemaker/`

Section 4: Advanced Topics

This section covers assorted topics on advanced usage of MLflow, allowing you to learn how to scale MLflow in big data environments and to meet high computing demands. We will cover advanced use cases to ensure broad coverage of the type of models and use cases to be discussed, alongside MLflow internals and how to contribute to it.

The following chapters are covered in this section:

- *Chapter 10, Scaling Up Your Machine Learning Workflow*
- *Chapter 11, Performance Monitoring*
- *Chapter 12, Advanced Topics with MLFlow*

10
Scaling Up Your Machine Learning Workflow

In this chapter, you will learn about diverse techniques and patterns to scale your **machine learning** (**ML**) workflow in different scalability dimensions. We will look at using a Databricks managed environment to scale your MLflow development capabilities, adding Apache Spark for cases where you have larger datasets. We will explore NVIDIA RAPIDS and **graphics processing unit** (**GPU**) support, and the Ray distributed frameworks to accelerate your ML workloads. The format of this chapter is a small **proof-of-concept** with a defined canonical dataset to demonstrate a technique and toolchain.

Specifically, we will look at the following sections in this chapter:

- Developing models with a Databricks Community Edition environment
- Integrating MLflow with Apache Spark
- Integrating MLflow with NVIDIA RAPIDS (GPU)
- Integrating MLflow with the Ray platform

This chapter will require researching the appropriate setup for each framework introduced, based on the standard official documentation for each of the cases.

Technical requirements

For this chapter, you will need the following prerequisites:

- The latest version of Docker installed on your machine. If you don't already have it installed, please follow the instructions at `https://docs.docker.com/get-docker/`.

- The latest version of Docker Compose installed—please follow the instructions at `https://docs.docker.com/compose/install/`.

- Access to Git in the command line, and installed as described in `https://git-scm.com/book/en/v2/Getting-Started-Installing-Git`.

- Access to a Bash terminal (Linux or Windows).

- Access to a browser.

- Python 3.5+ installed.

- The latest version of your ML library installed locally as described in *Chapter 3, Your Data Science Workbench*.

- An **Amazon Web Services** (**AWS**) account configured to run the MLflow model.

Developing models with a Databricks Community Edition environment

In many scenarios of small teams and companies, starting up a centralized ML environment might be a costly, resource-intensive, upfront investment. A team being able to quickly scale and getting a team up to speed is critical to unlocking the value of ML in an organization. The use of managed services is very relevant in these cases to start prototyping systems and to begin to understand the viability of using ML at a lower cost.

A very popular managed ML and data platform is the Databricks platform, developed by the same company that developed MLflow. We will use in this section the Databricks Community Edition version and license targeted for students and personal use.

In order to explore the Databricks platform to develop and share models, you need to execute the following steps:

1. Sign up to Databricks Community Edition at `https://community.cloud.databricks.com/` and create an account.

2. Log in to your account with your just-created credentials.

3. Upload training data into Databricks. You can start by uploading the training data available in the `Chapter10/databricks_notebooks/training_data.csv` folder. In the following screenshot, you can see represented the **Data** tab on the left, and you should see your file uploaded to the platform:

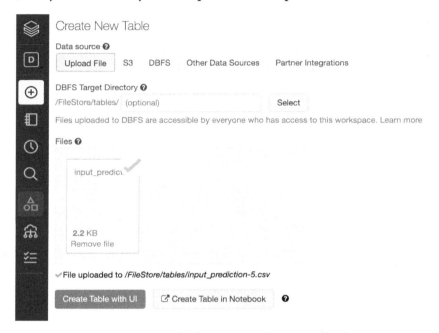

Figure 10.1 – Uploading training data to Databricks

4. Upload training data to Databricks. You can start by uploading the training data available in the `Chapter10/databricks_notebooks/input_prediction.csv` folder.

5. Create a cluster to use for your workloads. You are allowed to have clusters for your workloads with a limit of 15 **gigabytes** (**GB**) of **random-access memory** (**RAM**) and with usage for a defined period of time.

You can see an overview of the cluster-creation process in the following screenshot:

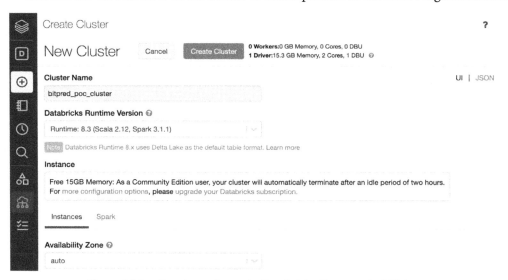

Figure 10.2 – Creating a cluster in Databricks Community Edition

6.   Create a new notebook in your Databricks platform on your landing workspace page by clicking on the **Create a Blank Notebook** button at the top right of the page, as illustrated in the following screenshot:

Figure 10.3 – Creating a new notebook in Databricks Community Edition

7. We are now ready to start a notebook to execute a basic training job in this managed environment. You can start by clicking on **Create Notebook**, as illustrated in the following screenshot:

Figure 10.4 – Creating your new notebook

8. Upload training data to Databricks. You can start by uploading the training data available in the Chapter10/databricks_notebooks/input_prediction.csv folder.

9. Import the needed libraries. We will adapt a LogicRegression model used to classify our running business case of the price of a btc-usd ticker, as follows:

```
import pandas
import numpy as np
import mlflow
from sklearn.linear_model import LogisticRegression
from sklearn.metrics import f1_score, confusion_matrix
from sklearn.model_selection import train_test_split
```

10. To read the data, due to the usage of the Databricks filesystem in the platform, it is more convenient to read the data in Spark and convert thereafter the DataFrame into `pandas`. We also split the data into training and test sets, as usual. Here is the code you'll need for this:

```
df = (spark.read.option("header","true").csv("/FileStore/
tables/training_data.csv"))
pandas_df = df.toPandas()
X=pandas_df.iloc[:,:-1]
Y=pandas_df.iloc[:,-1]
X_train, X_test, y_train, y_test = train_test_split(X, Y,
test_size=0.33, random_state=4284, stratify=Y)
```

11. Our next step will be to quickly train our classifier, as follows:

```
mlflow.sklearn.autolog()
model = LogisticRegression()
with mlflow.start_run(run_name='logistic_regression_
model_baseline') as run:
    model.fit(X_train, y_train)
    preds = model.predict(X_test)
```

12. In the top corner of the page, you can click on the **Experiment** button to view more details about your run, and you can click further to look at your model experiment, in the familiar interface of experiments, as illustrated in the following screenshot:

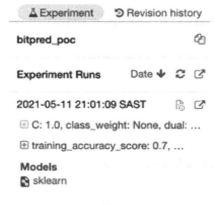

Figure 10.5 – Experiment button

13. One interesting feature that can scale and accelerate your ability to collaborate with others is the ability to publish model notebooks that are publicly accessible to everyone with whom you share a link, as illustrated in the following screenshot:

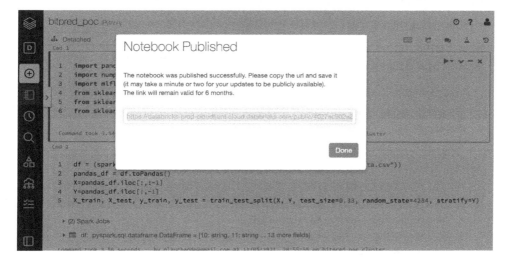

Figure 10.6 – Publishing notebooks

You can also export your notebook as a dbc file so that you can quickly start it up in a Databricks environment, and you can also share it in a repository, as you can see in the chapter folder, under /databricks-notebooks/bitpred_poc.dbc.

Having dealt with ways to scale your ability to run, develop, and distribute models using a Databricks environment, we will next look at integrating an Apache Spark flow into our inference workflows to handle scenarios where we have access to large datasets.

Integrating MLflow with Apache Spark

Apache Spark is a very scalable and popular big data framework that allows data processing at a large scale. For more details and documentation, please go to https://spark.apache.org/. As a big data tool, it can be used to speed up parts of your ML inference, as it can be set at a training or an inference level.

In this particular case, we will illustrate how to implement it to use the model developed in the previous section on the Databricks environment to scale the batch-inference job to larger amounts of data.

In other to explore Spark integration with MLflow, we will execute the following steps:

1. Create a new notebook named `inference_job_spark` in Python, linking to a running cluster where the `bitpred_poc.ipynb` notebook was just created.

2. Upload your data to `dbfs` on the File/Upload data link in the environment.

3. Execute the following script in a cell of the notebook, changing the `logged_model` and `df` filenames for the ones in your environment:

```python
import mlflow
logged_model = 'runs:/6815b44128e14df2b356c9db23b7f936/
model'



df = spark.read.format("csv").load("dbfs:/FileStore/
shared_uploads/ input.csv")
# Load model as a Spark UDF.
loaded_model = mlflow.pyfunc.spark_udf(spark, model_
uri=logged_model)


# Predict on a Spark DataFrame.
df.withColumn('predictions', loaded_model()).collect()
```

This illustrative excerpt running on Databricks or on your own Spark cluster can scale to large datasets, using the power of distributed computing in Spark.

From scaling inference with Apache Spark, we will look now at using GPUs with the support of MLflow to scale hyperparameter optimization jobs.

Integrating MLflow with NVIDIA RAPIDS (GPU)

Training and tuning ML models is a long and computationally expensive operation and is one of the operations that can benefit the most from parallel processing. We will explore in this section the integration of your MLflow training jobs, including hyperparameter optimization, with the NVIDIA RAPIDS framework.

To integrate the NVIDIA RAPIDS library, follow the next steps:

1. Install RAPIDS in the most convenient way for your environment, outlined as follows:

    a. `https://rapids.ai/start.html` contains detailed information on deployment options.

b. `https://developer.nvidia.com/blog/run-rapids-on-google-colab/` details how to run RAPIDS on **Google Colaboratory (Google Colab)**.

2. Install MLflow in your environment.

3. Import the needed libraries, as follows:

```
import argparse
from functools import partial

import mlflow
import mlflow.sklearn

from cuml.metrics.accuracy import accuracy_score
from cuml.preprocessing.model_selection import train_
test_split
from cuml.ensemble import RandomForestClassifier

from hyperopt import fmin, tpe, hp, Trials, STATUS_OK
```

4. Implement the `load_data` function, which is a helper function for loading data to be used by **central processing unit (CPU)**/GPU models. The `cudf` DataFrame is a DataFrame library for loading, joining, aggregating, and filtering without knowing the details of **Compute Unified Device Architecture (CUDA)** programming. Here is the code you'll need:

```
def load_data(fpath):
    import cudf
    df = cudf.read_parquet(fpath)
    X = df.drop(["ArrDelayBinary"], axis=1)
    y = df["ArrDelayBinary"].astype("int32")

    return train_test_split(X, y, test_size=0.2)Start the
ray server
ray.init()
client = serve.start()
```

5. Define a training loop, as follows:

```
def _train(params, fpath):
    max_depth, max_features, n_estimators = params
    max_depth, max_features, n_estimators = (int(max_
depth), float(max_features), int(n_estimators))
    X_train, X_test, y_train, y_test = load_data(fpath)
    mod = RandomForestClassifier(
        max_depth=max_depth, max_features=max_features,
n_estimators=n_estimators
    )
    mod.fit(X_train, y_train)
    preds = mod.predict(X_test)
    acc = accuracy_score(y_test, preds)

    mlparams = {
        "max_depth": str(max_depth),
        "max_features": str(max_features),
        "n_estimators": str(n_estimators),
    }

    mlflow.log_params(mlparams)
    mlflow.log_metric("accuracy", acc)
    mlflow.sklearn.log_model(mod, "saved_models")
    return {"loss": acc, "status": STATUS_OK}
```

6. Call the inner training loop, like this:

```
def train(params, fpath, hyperopt=False):

    with mlflow.start_run(nested=True):
        return _train(params, fpath, hyperopt)
```

7. Set up your main flow by reading an argument, if you are using the version deployed in Docker. The code to do this is illustrated in the following snippet:

```
if __name__ == "__main__":
    parser = argparse.ArgumentParser()
    parser.add_argument("--algo", default="tpe",
```

```
choices=["tpe"], type=str)
    parser.add_argument("--conda-env", required=True,
type=str)
    parser.add_argument("--fpath", required=True,
type=str)
    args = parser.parse_args()
```

8. Define your trials and parameters to optimize, as follows:

```
search_space = [
    hp.uniform("max_depth", 5, 20),
    hp.uniform("max_features", 0.1, 1.0),
    hp.uniform("n_estimators", 150, 1000),
]

trials = Trials()
algorithm = tpe.suggest if args.algo == "tpe" else
None
fn = partial(train, fpath=args.fpath, hyperopt=True)
experid = 0
```

9. Run your main loop, as follows:

```
artifact_path = "Airline-Demo"
artifact_uri = None


with mlflow.start_run(run_name="RAPIDS-Hyperopt"):
    argmin = fmin(fn=fn, space=search_space,
algo=algorithm, max_evals=2, trials=trials)


    print("============")
    fn = partial(train, fpath=args.fpath,
hyperopt=False)
    final_model = fn(tuple(argmin.values()))


    mlflow.sklearn.log_model(
        final_model,
        artifact_path=artifact_path,
```

```
                       registered_model_name="rapids_mlflow_cli",
                       conda_env="envs/conda.yaml",
            )
```

After having dealt with using a highly scalable compute environment to serve models on top of the Ray platform, we will now consider a different problem, where we will look at options to track multiple runs from a local machine in a centralized cloud location.

Integrating MLflow with the Ray platform

The Ray framework (`https://docs.ray.io/en/master/`) is a distributed platform that allows you to quickly scale the deployment infrastructure.

With Ray, you can add arbitrary logic when running an ML platform that needs to scale in the same way as model serving. It's basically a web framework.

We preloaded the model and contents that will be used into the following folder of the repository: `https://github.com/PacktPublishing/Machine-Learning-Engineering-with-MLflow/tree/master/Chapter10/mlflow-ray-serve-integration`.

In order to execute your model serving into Ray, execute the following steps:

1. Install the Ray package by running the following command:

    ```
    pip install -U ray
    ```

2. Install MLflow in your environment.

3. Import the needed libraries, as follows:

    ```
    import ray
    from ray import serve

    import mlflow.pyfunc
    ```

4. Implement the model backend, which basically means wrapping up the model-serving function into your Ray serving environment. Here's the code you'll need:

    ```
    class MLflowBackend:
        def __init__(self, model_uri):
            self.model = mlflow.pyfunc.load_model(model_
    ```

```
        uri=model_uri)

        async def __call__(self, request):
            return self.model.predict(request.data)
```

5. Start the Ray server, as follows:

```
ray.init()
client = serve.start()
```

6. Load the model and create a backend, like this:

```
model_uri = "./tmp/0/31fc9974587243d181fdbebfd4d2b6ad/
artifacts/model"
client.create_backend("mlflow_backend", MLflowBackend,
model_uri)
```

7. Test the serving platform by running the following command:

```
ray start --head # Start local Ray cluster.
serve start # Start Serve on the local Ray cluster.
```

After having dealt with using a highly scalable compute environment to serve models on top of the Ray platform, we will look at the performance and monitoring component in the following chapter.

Summary

In this chapter, we focused on scaling your ability to run, develop, and distribute models using a Databricks environment. We also looked at integrating an Apache Spark flow into our batch-inference workflows to handle scenarios where we have access to large datasets.

We concluded the chapter with two approaches to scale hyperparameter optimization and **application programming interface** (**API**) serving with scalability, using the NVIDIA RAPIDS framework and the Ray distributed framework.

In the next chapter and in further sections of the book, we will focus on the observability and performance monitoring of ML models.

Further reading

In order to further your knowledge, you can consult the documentation at the following links:

- `https://www.mlflow.org/docs/latest/python_api/mlflow.sagemaker.html`

- `https://aws.amazon.com/blogs/machine-learning/managing-your-machine-learning-lifecycle-with-mlflow-and-amazon-sagemaker/`

- `https://docs.databricks.com/applications/mlflow/index.html`

11
Performance Monitoring

In this chapter, you will learn about the important and relevant area of **Machine Learning (ML)** operations and how to ensure a smooth ride in the production systems developed so far in this book using best practices in the area and known operational patterns. We will understand the concept of operations in ML, and look at metrics for monitoring data quality in ML systems.

Specifically, we will look at the following sections in this chapter:

- Overview of performance monitoring for ML models
- Monitoring data drift and model performance
- Monitoring target drift
- Infrastructure monitoring and alerting

We will address some practical reference tools for performance and reliability monitoring of ML systems.

Technical requirements

For this chapter, you will need the following prerequisites:

- The latest version of Docker installed on your machine. If you don't already have it installed, please follow the instructions at `https://docs.docker.com/get-docker/`.

- The latest version of `docker-compose` installed. To do this, please follow the instructions at `https://docs.docker.com/compose/install/`.

- Access to Git in the command line, which can be installed as described at `https://git-scm.com/book/en/v2/Getting-Started-Installing-Git`.

- Access to a Bash terminal (Linux or Windows).

- Access to a browser.

- Python 3.8+ installed.

- The latest version of your ML platform installed locally as described in *Chapter 3, Your Data Science Workbench*.

- An AWS account configured to run the MLflow model.

Overview of performance monitoring for machine learning models

Monitoring is at the cornerstone of reliable ML systems able to consistently unlock the value of data and provide critical feedback for improvement.

On the monitoring side of ML models, there are multiple interested parties, and we should take the requirements for monitoring from the different stakeholders involved. One example of a typical set of stakeholders is the following:

- **Data scientists**: Their focus regarding monitoring is evaluating model performance and data drift that might negatively affect that performance.

- **Software engineers**: These stakeholders want to ensure that they have metrics that assess whether their products have reliable and correct access to the APIs that are serving models.

- **Data engineers**: They want to ensure that the data pipelines are reliable and pushing data reliably, at the right velocity, and in line with the correct schemas.

- **Business/product stakeholders**: These stakeholders are interested in the core impact of the overall solution on their customer base. For instance, in a trading platform, they might be most concerned with the profit-to-risk ratio that the overall solution brings to the company. A circuit breaker might be added to the algorithm if the market is in a day of very high volatility or in an atypical situation.

The most widely used dimensions of monitoring in the ML industry are the following:

- **Data drift**: This corresponds to significant changes in the input data used either for training or inference in a model. It might indicate a change of the modeled premise in the real world, which will require the model to be retrained, redeveloped, or even archived if it's no longer suitable. This can be easily detected by monitoring the distributions of data used for training the model versus the data used for scoring or inference over time.

- **Target drift**: In line with the change of regimens in input data, we often see the same change in the distribution of outcomes of the model over a period of time. The common periods are months, weeks, or days, and might indicate a significant change in the environment that would require model redevelopment and tweaking.

- **Performance drift**: This involves looking at whether the performance metrics such as accuracy for classification problems, or root mean square error, start suffering a gradually worsening over time. This is an indication of an issue with the model requiring investigation and action from the model developer or maintainer.

- **Platform and infrastructure metrics**: This type of metrics is not directly related to modeling, but with the systems infrastructure that encloses the model. It implies abnormal CPU, memory, network, or disk usage that will certainly affect the ability of the model to deliver value to the business.

- **Business metrics**: Very critical business metrics, such as the profitability of the models, in some circumstances should be added to the model operations in order to ensure that the team responsible for the model can monitor the ability of the model to deliver on its business premise.

In the next section, we will look at using a tool that we can integrate with **MLflow** to monitor for data drift and check the performance of models.

Monitoring data drift and model performance

In this section, we will run through an example that you can follow in the notebook available in the **GitHub** repository (at `https://github.com/PacktPublishing/Machine-Learning-Engineering-with-MLflow/tree/master/Chapter11/model_performance_drifts`) of the code of the package. We will run through the process of calculating different types of drift and exploring its integration with MLflow.

One emergent open source tool in the space of monitoring model performance is called **Evidently** (`https://evidentlyai.com/`). Evidently aids us in analyzing ML models during the production and validation phases. It generates handy reports integrated with `pandas`, JSON, and CSV. It allows us to monitor multiple drifts in ML models and their performance. The GitHub repository for Evidently is available at `https://github.com/evidentlyai/evidently/`.

In this section, we will explore the combination of Evidently with MLflow, in order to monitor data drift and model performances in the next section.

Monitoring data drift

In this subsection, we will set up **Evidently** in our environment and understand how to integrate it. Follow these steps in the GitHub repository (refer to the *Technical requirements* section for more details):

1. Install `evidently`:

    ```
    pip install evidently==0.1.17.dev0
    ```

2. Import the relevant libraries:

    ```
    import pandas as pd
    import numpy as np
    from sklearn import datasets
    from sklearn.model_selection import train_test_split
    from evidently.dashboard import Dashboard
    from evidently.tabs import DataDriftTab,
    NumTargetDriftTab,CatTargetDriftTab
    ```

3. Get a reference dataset, basically a training dataset. We will add a set of features to the `pandas` DataFrame so `evidently` will be able to use the feature names in the drift reports:

```
reference_data = \
pd.read_csv("training_data.csv", header=None,
        names=[ "day{}".format(i) for i in \
            range(0,14) ]+["target"] )
```

The following *Figure 11.1* represents the data structure of the training data that we will be using as the reference dataset:

	day0	day1	day2	day3	day4	day5	day6	day7	day8	day9	day10	day11	day12	day13	target
0	1	1	0	0	0	0	1	1	0	0	0	1	0	0	0
1	1	0	0	0	0	1	1	0	0	0	1	0	0	0	1
2	0	0	0	0	1	1	0	0	0	1	0	0	0	1	0
3	0	0	0	1	1	0	0	0	1	0	0	0	1	0	1
4	0	0	1	1	0	0	0	1	0	0	0	1	0	1	1

Figure 11.1 – Sample of the dataset to be used

4. In this step, we load the `to_score_input_data.csv` file. This is the file to be scored. Our intention later in this exercise is to calculate the distribution difference between the data in the reference training set and the data to be scored:

```
latest_input_data = \
pd.read_csv("to_score_input_data.csv", header=None,
        names=[ "day{}".format(i) for i in \
            range(0,14) ] )
```

5. Execute the data drift report generation and log into an MLflow run. Basically, what happens in the following code excerpt is the generation of an Evidently dashboard with the reference data and the latest input data. A drift report is calculated and loaded into an MLflow run so it can be actioned and reviewed in further steps:

```
EXPERIMENT_NAME="./reports_data_drift"
mlflow.set_experiment(EXPERIMENT_NAME)
with mlflow.start_run():
    drift_dashboard = Dashboard(tabs=[DataDriftTab])
    drift_dashboard.calculate(reference_data,
```

```
                                          latest_input_data)
    drift_dashboard.save(EXPERIMENT_NAME+"/input_data_
drift.html")
    drift_dashboard._save_to_json(EXPERIMENT_NAME+"/
input_data_drift.json")
    mlflow.log_artifacts(EXPERIMENT_NAME)
```

6. You can run now the notebook code (on the `monitoring_data_drift_performance.ipynb` file) of the previous cells and explore your data drift reports in the MLflow UI over the Artifacts component of the MLflow run. *Figure 11.2* shows that the tool didn't detect any drift among the 14 features, and the distributions are presented accordingly:

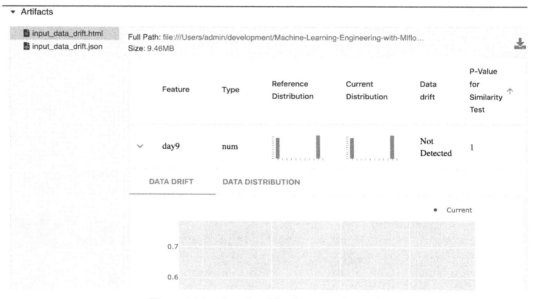

Figure 11.2 – Sample of the dataset to be used

In a similar fashion to data drift, we will now look in the next subsection at target drift to uncover other possible issues in our model.

Monitoring target drift

We will now compare the scored output with the reference training output to look for possible target drift:

1. Get the recently scored dataset:

```
production_scored_data = \
pd.read_csv("scored_data.csv", header=None,
            names=[ "day{}".format(i) for i in \
                    range(0,14) ]+["target"] )


bcancer_data_and_target_drift = \
Dashboard(reference_data, production_scored_data,
        tabs=[ CatTargetDriftTab])


bcancer_data_and_target_drift.save('reports/target_drift.
html')
```

2. Execute the data drift report generation and log the results in MLflow:

```
EXPERIMENT_NAME="./reports_target_drift"
mlflow.set_experiment(EXPERIMENT_NAME)
with mlflow.start_run():
    model_target_drift = \
    Dashboard(reference_data, production_scored_data,
            tabs=[CatTargetDriftTab])
    model_target_drift.save(EXPERIMENT_NAME+"/target_
drift.html")
    drift_dashboard._save_to_json(EXPERIMENT_NAME+"/
target_drift.json")
    mlflow.log_artifacts(EXPERIMENT_NAME)
```

3. Explore the target drift reports on your target. As can be seen in *Figure 11.3*, no statistically significant figure on this run was found for target drift. In detecting drift, Evidently does statistical tests using the probability of the data being from a different distribution represented by the **p-value** (more details on this can be found at `https://en.wikipedia.org/wiki/P-value`). It compares the results between the reference and the current data:

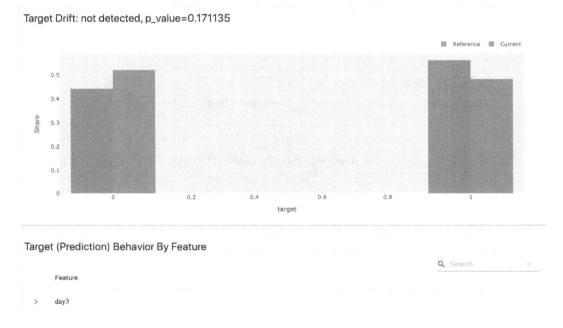

Target Drift: not detected, p_value=0.171135

Target (Prediction) Behavior By Feature

Figure 11.3 – Target data drift for target

4. As shown in *Figure 11.4*, you can drill down further into target drift on a specific feature; in this case, a specific previous **day8** to predict the stock price:

▾ Artifacts

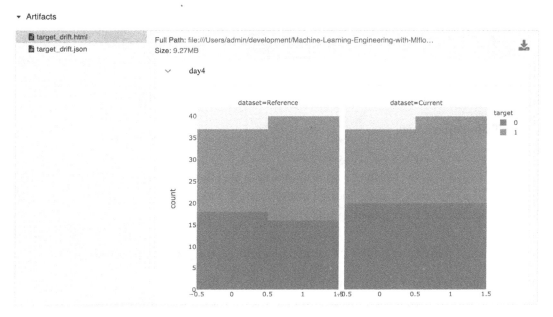

Figure 11.4 – Target data drift for our target

After having learned how to detect drift in the input data, we will now look at how to use Evidently to monitor drift in models.

Monitoring model drift

Monitoring model drift is extremely important to ensure that your model is still delivering at its optimal performance level. From this analysis, you can make a decision on whether to retrain your model or even develop a new one from scratch.

We will now monitor model drift. To do this, you need to execute the following steps:

1. Import the relevant libraries:

```
import xgboost as xgb
import mlflow
from evidently.tabs import ClassificationPerformanceTab
```

2. Get a reference dataset:

```
X=reference_data.iloc[:,:-1]
Y=reference_data.iloc[:,-1]

reference, production, y_train, y_test = \
```

```
train_test_split(X, Y, test_size=0.33,
                 random_state=4284, stratify=Y)
reference_train = xgb.DMatrix(reference,label=y_train)
dproduction= xgb.DMatrix(production)
dreference=xgb.DMatrix(reference)
```

3. Train your model:

```
mlflow.xgboost.autolog()
EXPERIMENT_NAME="reports_model_performance"
mlflow.set_experiment(EXPERIMENT_NAME)
with mlflow.start_run() as run:

    model=xgb.train(dtrain=reference_train,params={})
```

4. Create a reference prediction and training predictions:

```
    train_proba_predict = model.predict(dreference)
    test_proba_predict = model.predict(dproduction)
    test_predictions = [1. if y_cont > threshold else 0.
for y_cont in test_proba_predict]
    train_predictions = [1. if y_cont > threshold else 0.
for y_cont in train_proba_predict]
    reference['target'] = y_train
    reference['prediction'] = train_predictions
    production['target'] = y_test
    production['prediction'] = test_predictions
```

5. Generate and attach the performance reports to your execution:

```
    classification_performance = Dashboard(
                    tabs=[ClassificationPerformanceTab])
    classification_performance.calculate(reference,
                                            production)
    classification_performance.save('.
reports/'+EXPERIMENT_NAME+'.html')
    mlflow.log_artifact('.reports/'+EXPERIMENT_NAME+'.
html')
```

6. Explore your MLflow performance metrics report. By looking at the reports generated, you can check on the **Reference** metrics that **Accuracy**, **Precision**, **Recall**, and **F1 metrics**, which are considered the reference metrics based on the training data, have maximum values of **1**. The current status on the row below is definitely degraded when we test the subset of testing data. This can help you make the call on whether it is sensible for the model to still be in production with the current **F1** value:

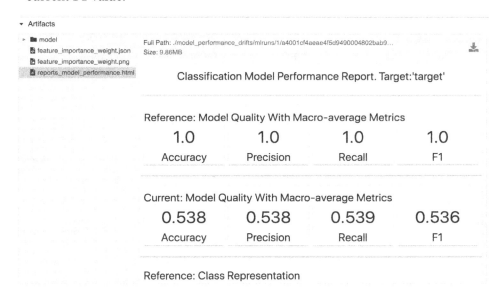

Figure 11.5 – Target data drift for target

After having delved into the details of data drift, target drift, and model performance monitoring, along with how to integrate these functionalities with MLflow, we will now look at the basic principles of monitoring infrastructure, including monitoring and alerting.

Infrastructure monitoring and alerting

The main dimensions of monitoring in ML systems from an infrastructure perspective do not differ from those in traditional software systems.

In order to illustrate this exact issue, we will leverage the monitoring and alerting tools available in **AWS CloudWatch** and **SageMaker** to illustrate an example of setting up monitoring and alerting infrastructure. This same mechanism can be set up with tools such as Grafana/Prometheus for on-premises and cloud deployments alike. These monitoring tools achieve similar goals and provide comparable features, so you should choose the most appropriate depending on your environment and cloud provider.

AWS CloudWatch provides a monitoring and observability solution. It allows you to monitor your applications, respond to system-wide performance changes, optimize resource use, and receive a single view of operational health.

At a higher level, we can split the infrastructure monitoring and alerting components into the following three items:

- **Resource metrics**: This refers to metrics regarding the hardware infrastructure where the system is deployed. The main metrics in this case would be the following:

  a. **CPU utilization**: This is basically a unit of utilization of your processor as a percentage value. This is the general metric available and should be monitored.

  b. **Memory utilization**: The percentage of memory in use at the moment by your computing system.

  c. **Network data transfer**: Network data transfer refers to the amount of traffic in and out of a specific compute node. It is generally measured in Mb/s. An anomaly might mean that you need to add more nodes to your system or increase capacity.

  d. **Disk I/O**: This is measured in the throughput of writes and reads from the disk; it might point to a system under stress that needs to be either scaled or have its performance investigated:

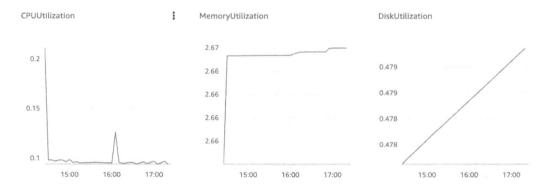

Figure 11.6 – SageMaker infrastructure metric examples

- **System metrics**: The second pillar of infrastructure monitoring and alerting components refers to metrics regarding the system infrastructure where the system is deployed. The main metrics in this case would be the following:

  a. **Request throughput**: The number of predictions served over a second

  b. **Error rate**: The number of errors per prediction

  c. **Request latencies**: The end-to-end time taken to serve a prediction

  d. **Validation metrics**: Error metrics on input data for the request

A production system such as SageMaker pushes system metrics into AWS CloudWatch to provide real-time system metrics monitoring. AWS CloudWatch has a complete feature set of features to manage, store, and monitor metrics and dashboards:

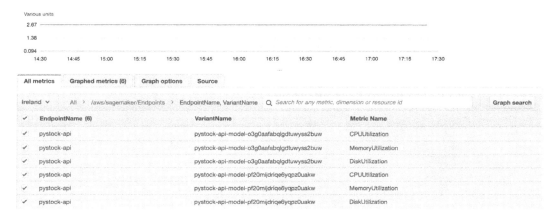

Figure 11.7 – Specify an alarm in AWS CloudWatch

- **Alerting**: For alerting, we use any of the metrics calculated in the previous section and set up a threshold that we consider acceptable. The AWS CloudWatch interface allows you to easily set up alerts on the default service metrics and custom metrics. The team responsible for reliability is alerted by CloudWatch sending messages to a corporate chat/Slack, email address, or mobile phone to allow the team to address or mitigate the incident:

Figure 11.8 – Specify an alarm in AWS CloudWatch

You can use the same monitoring tools to log and monitor all the other metrics that are interrelated with your ML systems. For instance, having an alert for the weekly profit of a ML model is a business metric that should be deployed alongside the core systems metrics of your system.

After being exposed to an overview of AWS CloudWatch as an example of a tool to implement metrics monitoring and alerting for your ML systems in production, we will explore advanced concepts of MLflow in the last chapter of the book.

Summary

In this chapter, we introduced the concepts of data drift and target drift, and examined different approaches to performance monitoring in ML systems.

We started by introducing important concepts in the realm of performance and monitoring, different types of drift and business metrics to monitor, and the use of AWS CloudWatch as a tool to implement monitoring and alerting in real-time systems.

Performance and monitoring is an important component of our architecture, and it will allow us to conclude an important layer of our ML system's architecture. Now let's delve into the next chapter on advanced topics in MLflow.

Further reading

In order to further your knowledge, you can consult the documentation at the following links:

- `https://www.mlflow.org/docs/latest/projects.html`
- `https://evidentlyai.com/`
- `https://aws.amazon.com/cloudwatch/`

12
Advanced Topics with MLflow

In this chapter, we will cover advanced topics to address common situations and use cases whereby you can leverage your MLflow knowledge by using different types of models from the ones exposed in the rest of the book, to ensure a breadth of feature coverage and exposure to assorted topics.

Specifically, we will look at the following sections in this chapter:

- Exploring MLflow use cases with AutoML
- Intergrating MLflow with other languages
- Understanding MLflow plugins

We will represent each of the cases with a brief description of the problem and solutions in a pattern format—namely, a problem context and a solution approach.

The different sections of this chapter don't present continuity as they address different issues.

Technical requirements

For this chapter, you will need the following prerequisites:

- The latest version of Docker installed on your machine. If you don't already have it installed, please follow the instructions at `https://docs.docker.com/get-docker/`.

- The latest version of Docker Compose installed—please follow the instructions at `https://docs.docker.com/compose/install/`.

- Access to Git in the command line, and installed as described at `https://git-scm.com/book/en/v2/Getting-Started-Installing-Git`.

- Access to a Bash terminal (Linux or Windows).

- Access to a browser.

- Python 3.5+ installed.

- The latest version of your ML library installed locally as described in *Chapter 4, Experiment Management in MLflow*.

Exploring MLflow use cases with AutoML

Executing an ML project requires a breadth of knowledge in multiple areas and, in a lot of cases, deep technical steps of expertise. One emergent technique to ease the adoption and accelerate **time to market** (**TTM**) in projects is the use of **automated machine learning** (**AutoML**), where some of the activities of the model developer are automated. It basically consists of automating steps in ML in a twofold approach, outlined as follows:

- **Feature selection**: Using optimization techniques (for example, Bayesian techniques) to select the best features as input to a model

- **Modeling**: Automatically identifying a set of models to use by testing multiple algorithms using hyperparameter optimization techniques

We will explore the integration of MLflow with an ML library called PyCaret (https://pycaret.org/) that allows us to leverage its AutoML techniques and log the process in MLflow so that you can automatically obtain the best performance for your problem.

We will look next at the use case of pyStock in the book and will look at automatically modeling based on our training data.

AutoML pyStock classification use case

For this section, we will work on a solution that you can follow along with (https://github.com/PacktPublishing/Machine-Learning-Engineering-with-MLflow/tree/master/Chapter12/automl_pycaret) with the notebook and our project dataset. We will execute the following steps in order to implement AutoML for our use case:

1. Let's start by installing the full version of PyCaret, as follows:

    ```
    pip install pycaret==2.3.1
    ```

2. First, we should import the necessary libraries, like so:

    ```
    import pandas
    import pycaret
    ```

3. Then, we read all the training data, like this:

    ```
    data=pandas.read_csv("training_data.csv",header='infer')
    ```

4. Next, we set up the project data and load the input data, as follows:

```
from pycaret.classification import *
s = setup(data, target = 'target',  log_experiment =
True, experiment_name = 'psystock')
```

Here is the output:

Following data types have	
	Data Type
day0	Categorical
day1	Categorical
day2	Categorical
day3	Categorical
day4	Categorical
day5	Categorical
day6	Categorical
day7	Categorical
day8	Categorical
day9	Categorical
day10	Categorical
day11	Categorical
day12	Categorical
day13	Categorical
target	Label

Figure 12.1 – Automatic feature inference

5. Then, we execute `compare_models()`, like this:

```
best = compare_models()
```

Here is the output:

	Model	Accuracy	AUC	Recall	Prec.	F1	Kappa	MCC	TT (Sec)
dt	Decision Tree Classifier	0.6133	0.6083	0.6333	0.6683	0.6212	0.2126	0.2314	0.0050
lr	Logistic Regression	0.5733	0.6167	0.6667	0.6367	0.6281	0.1186	0.1474	0.3750
nb	Naïve Bayes	0.5267	0.5667	0.6000	0.5667	0.5752	0.0270	0.0391	0.0060
lightgbm	Light Gradient Boosting Machine	0.5233	0.5778	0.6000	0.6117	0.5695	0.0186	0.0507	0.0410
knn	K Neighbors Classifier	0.5200	0.4556	0.7000	0.5367	0.6024	-0.0226	-0.0145	0.0080
ada	Ada Boost Classifier	0.5167	0.6111	0.5333	0.5683	0.5055	0.0538	0.0789	0.0330
xgboost	Extreme Gradient Boosting	0.5033	0.5278	0.5667	0.5333	0.5419	-0.0201	-0.0255	0.0910
ridge	Ridge Classifier	0.4967	0.0000	0.5333	0.5533	0.4948	-0.0008	0.0176	0.0050
qda	Quadratic Discriminant Analysis	0.4967	0.4722	0.6000	0.5250	0.5450	-0.0285	-0.0522	0.0060
lda	Linear Discriminant Analysis	0.4967	0.6167	0.5333	0.5533	0.4948	-0.0008	0.0176	0.0060
catboost	CatBoost Classifier	0.4500	0.3667	0.5333	0.4750	0.4943	-0.1054	-0.1242	0.7460
gbc	Gradient Boosting Classifier	0.4467	0.4944	0.5667	0.4500	0.5000	-0.1321	-0.1537	0.0240
rf	Random Forest Classifier	0.4333	0.3917	0.5333	0.4417	0.4640	-0.1281	-0.1614	0.0880
svm	SVM - Linear Kernel	0.4300	0.0000	0.4667	0.4200	0.4038	-0.1594	-0.1742	0.0050
et	Extra Trees Classifier	0.4300	0.4389	0.5333	0.4533	0.4748	-0.1767	-0.2021	0.0770

Figure 12.2 – Different types of models

6. Select your best model by running the following command:

```
best = compare_models()
```

7. Run MLflow to check all the models (in the following **Uniform Resource Locator (URL)**: `http://127.0.0.1:5000/#/experiments/1`), and you should then see a screen like this:

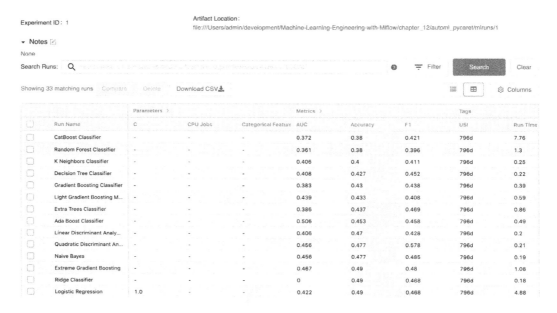

Experiment ID: 1

Artifact Location:
file:///Users/admin/development/Machine-Learning-Engineering-with-Mlflow/chapter_12/automl_pycaret/mlruns/1

▾ Notes ✎
None

Search Runs: Q   ❔   ☰ Filter   [Search]   Clear

Showing 33 matching runs   Compare   Delete   Download CSV⬇   ☰ ⊞ ⚙ Columns

	Run Name	Parameters >			Metrics >			Tags	
		C	CPU Jobs	Categorical Feature	AUC	Accuracy	F1	USI	Run Time
☐	CatBoost Classifier	-	-	-	0.372	0.38	0.421	796d	7.76
☐	Random Forest Classifier	-	-	-	0.361	0.38	0.396	796d	1.3
☐	K Neighbors Classifier	-	-	-	0.406	0.4	0.411	796d	0.25
☐	Decision Tree Classifier	-	-	-	0.408	0.427	0.452	796d	0.22
☐	Gradient Boosting Classifier	-	-	-	0.383	0.43	0.438	796d	0.39
☐	Light Gradient Boosting M...	-	-	-	0.439	0.433	0.408	796d	0.59
☐	Extra Trees Classifier	-	-	-	0.386	0.437	0.469	796d	0.86
☐	Ada Boost Classifier	-	-	-	0.506	0.453	0.458	796d	0.49
☐	Linear Discriminant Analy...	-	-	-	0.406	0.47	0.428	796d	0.2
☐	Quadratic Discriminant An...	-	-	-	0.456	0.477	0.578	796d	0.21
☐	Naïve Bayes	-	-	-	0.456	0.477	0.485	796d	0.19
☐	Extreme Gradient Boosting	-	-	-	0.467	0.49	0.48	796d	1.08
☐	Ridge Classifier	-	-	-	0	0.49	0.468	796d	0.18
☐	Logistic Regression	1.0	-	-	0.422	0.49	0.468	796d	4.88

Figure 12.3 – Models logged in MLflow

We will next look at implementing AutoML in a scenario where we don't have targets. We will need to use anomaly detection, a non-supervised ML technique.

AutoML – anomaly detection in fraud

For this section, we will work on a solution that you can follow along with (`https://github.com/PacktPublishing/Machine-Learning-Engineering-with-MLflow/tree/master/Chapter12/automl_pycaret_fraud`) with the notebook and our project dataset. We will execute the following steps in order to implement AutoML for our use case:

1. First, we should import the libraries, like so:

```
import pandas
import pycaret
```

2. Then, we read all the training data, like this:

```
data=pandas.read_csv("credit_card.csv",header='infer')
```

Here is the output:

V9	...	V20	V21	V22	V23	V24	V25	V26	V27	V28	Amount
0.363787	...	0.251412	-0.018307	0.277838	-0.110474	0.066928	0.128539	-0.189115	0.133558	-0.021053	149.62
-0.255425	...	-0.069083	-0.225775	-0.638672	0.101288	-0.339846	0.167170	0.125895	-0.008983	0.014724	2.69
-1.514654	...	0.524980	0.247998	0.771679	0.909412	-0.689281	-0.327642	-0.139097	-0.055353	-0.059752	378.66
-1.387024	...	-0.208038	-0.108300	0.005274	-0.190321	-1.175575	0.647376	-0.221929	0.062723	0.061458	123.50
0.817739	...	0.408542	-0.009431	0.798278	-0.137458	0.141267	-0.206010	0.502292	0.219422	0.215153	69.99
...	...	...	...	...	...	...	...	...	...	...	...
1.914428	...	1.475829	0.213454	0.111864	1.014480	-0.509348	1.436807	0.250034	0.943651	0.823731	0.77
0.584800	...	0.059616	0.214205	0.924384	0.012463	-1.016226	-0.606624	-0.395255	0.068472	-0.053527	24.79
0.432454	...	0.001396	0.232045	0.578229	-0.037501	0.640134	0.265745	-0.087371	0.004455	-0.026561	67.88
0.392087	...	0.127434	0.265245	0.800049	-0.163298	0.123205	-0.569159	0.546668	0.108821	0.104533	10.00
0.486180	...	0.382948	0.261057	0.643078	0.376777	0.008797	-0.473649	-0.818267	-0.002415	0.013649	217.00

Figure 12.4 – Models automatically available in MLflow

3. Next, we set up the project data and load the input data, as follows:

```
from pycaret.anomaly import *
s = setup(df,  log_experiment = True, experiment_name =
'psystock_anomaly'))
```

4. Then, we execute compare_models(), like this:

```
models()
```

Here is the output:

ID	Name	Reference
abod	Angle-base Outlier Detection	pyod.models.abod.ABOD
cluster	Clustering-Based Local Outlier	pyod.models.cblof.CBLOF
cof	Connectivity-Based Local Outlier	pyod.models.cof.COF
iforest	Isolation Forest	pyod.models.iforest.IForest
histogram	Histogram-based Outlier Detection	pyod.models.hbos.HBOS
knn	K-Nearest Neighbors Detector	pyod.models.knn.KNN
lof	Local Outlier Factor	pyod.models.lof.LOF
svm	One-class SVM detector	pyod.models.ocsvm.OCSVM
pca	Principal Component Analysis	pyod.models.pca.PCA
mcd	Minimum Covariance Determinant	pyod.models.mcd.MCD
sod	Subspace Outlier Detection	pyod.models.sod.SOD
sos	Stochastic Outlier Selection	pyod.models.sos.SOS

Figure 12.5 – Different types of models

5. Then, execute your chosen anomaly detection model, as follows:

```
iforest = create_model('iforest', fraction = 0.1)
iforest_results = assign_model(iforest)
iforest_results.head()
```

6. Next, run MLflow to check all the models (at the following URL: `http://127.0.0.1:5000/#/experiments/1`), and you should see a screen like this:

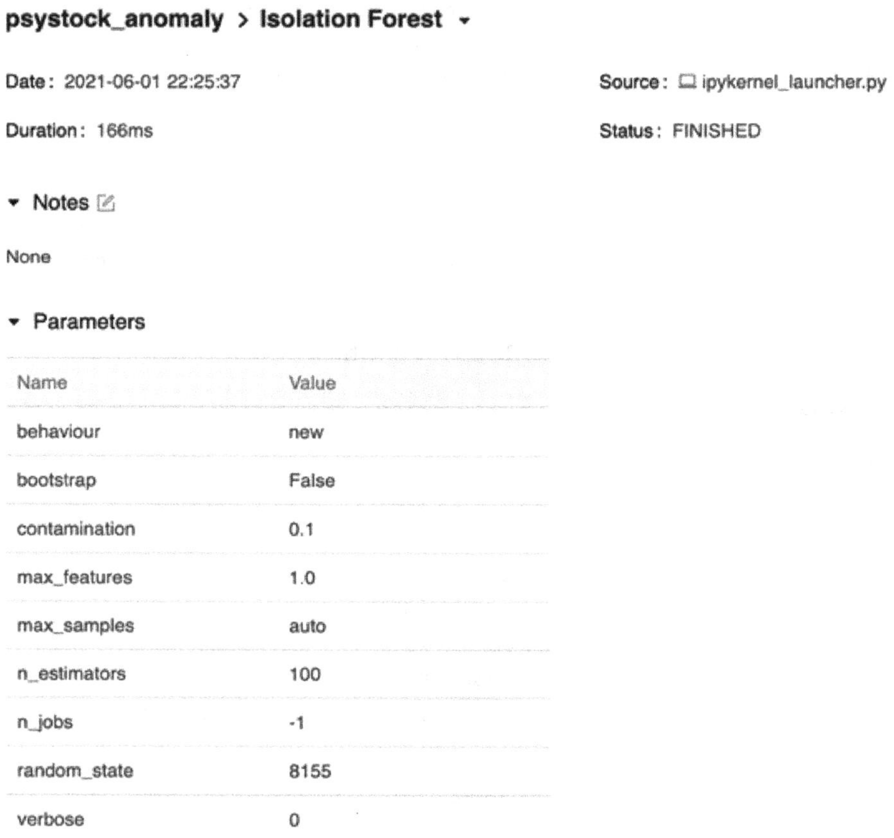

psystock_anomaly > Isolation Forest ▾

Date: 2021-06-01 22:25:37   Source: ▭ ipykernel_launcher.py

Duration: 166ms   Status: FINISHED

▾ Notes ☑

None

▾ Parameters

Name	Value
behaviour	new
bootstrap	False
contamination	0.1
max_features	1.0
max_samples	auto
n_estimators	100
n_jobs	-1
random_state	8155
verbose	0

Figure 12.6 – Models automatically available in MLflow

At this stage, you should be able to leverage the knowledge you have gained throughout the book to use the models identified in this book for models in production. We will next look at intergrating MLflow with other languages—in this case, Java.

Integrating MLflow with other languages

MLflow is primarily a tool ingrained in the Python ecosystem in the ML space. At its core, MLflow components provide a **REpresentational State Transfer** (**REST**) interface. As long as **application programming interface** (**API**) wrappers are made, the underlying code is accessible from any language with REST support. The REST interface is extensively documented in `https://www.mlflow.org/docs/latest/rest-api.html`; most of the integration into other languages is about providing layers to access the API in a concise, language-specific library.

MLflow Java example

Multiple teams in the ML space are inserted in a context where multiple languages are used. One of the most important platforms on large-scale distributed systems is **Java Virtual Machine** (**JVM**). Being able to implement systems that can interact with Java-based systems is paramount for a smooth integration of MLflow with the wider **information technology** (**IT**) infrastructure.

We will show an example of using MLflow in Java (you can have access to the code here: `https://github.com/PacktPublishing/Machine-Learning-Engineering-with-MLflow/tree/master/Chapter12/psystock-java-example`). In order to use MLflow in Java, you will have to execute the following steps:

1. Install Java and the Java build tool called `Maven`, as directed by `https://maven.apache.org/install.html`.

2. Create a dependencies `pom.xml` file with the MLflow client dependency, as follows:

```
<project>
...
  <dependencies>
    <dependency>
       <groupId>org.mlflow</groupId>
       <artifactId>mlflow-client</artifactId>
       <version>1.17.0</version>..
    </dependency>
  ...
</project>
```

3. Implement your main class, like this:

```
package ai.psystock.jclient;

import org.mlflow.tracking.MlflowClient;
import org.mlflow.tracking.MlflowContext;
import java.io.File;
import java.io.PrintWriter;

public class Main {
    public static void main(String[] args) {

        MlflowClient mlflowClient=new MlflowClient();
        String runId="test";
        RunStatus = RunStatus.FINISHED;

        MlflowContext = new MlflowContext();
        MlflowClient client = mlflowContext.getClient();

        client.logParam("test","alpha", "0.5");
        client.logMetric("test","rmse", 0.786);
        client.setTag("test","origin","HelloWorldFluent
Java Example");

        mlflowClient.setTerminated(runId, runStatus,
System.currentTimeMillis());
    }
}
```

4. Build your project with Maven, as follows:

```
mvn clean package
```

5. Execute your Java project by running the following code:

```
java -jar ./target/java-maven-command-line-1.0-SNAPSHOT.
jar
```

At this stage, MLflow is natively integrated into the Python ecosystem. It provides links to other ecosystems similar to what we demonstrated in this chapter with the JVM language.

We will next explore an example in the R language.

MLflow R example

We will show an example of using MLflow in R using the Databricks environment (you can have access to the code here: `https://github.com/PacktPublishing/Machine-Learning-Engineering-with-MLflow/tree/master/Chapter12/mlflow-example-r`). You can import the notebook from the Databricks Community Edition environment and explore the code from there.

In this section, we will run a random forest classifier in R over the standard dataset available as an R package, called `Pima.tf` (`https://rdrr.io/cran/MASS/man/Pima.tr.html`). This is a simple dataset with a set of biomedical features to detect whether a specific patient has diabetes or not.

In order to create a notebook for your R example code, you need to execute the following steps:

1. Sign up to Databricks Community Edition at `https://community.cloud.databricks.com/` and create an account.

2. Log in to your account with your just-created credentials.

3. Create a cluster to use for your workloads. You are allowed to have clusters for your workloads with a limit of 15 **gigabytes (GB)** of **random-access memory (RAM)** and with usage for a defined period of time.

    You can see an overview of the cluster-creation process in the following screenshot:

Figure 12.7 – Creating a cluster in Databricks Community Edition

4. Create a new notebook in your Databricks platform on your landing workspace page by clicking on the **Create a Blank Notebook** button in the top right of the page, as illustrated in the following screenshot:

Figure 12.8 – Creating a new notebook in Databricks Community Edition

5. We are now ready to start a notebook to execute a basic training job in this managed environment. You can start by clicking on **New Notebook** in your workspace. You need to set the default language as **R** and attach the notebook to your cluster created in the previous chapter.

    You can see an overview of the notebook-creation process in the following screenshot:

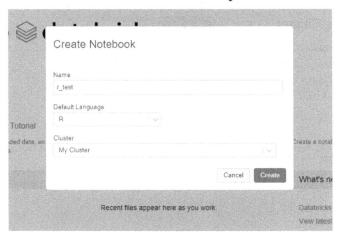

Figure 12.9 – Adding details of your new R notebook

6. You start on your notebook by importing the MLflow dependencies through `install.packages` and by instantiating the library, as follows:

```
install.packages("mlflow")
library(mlflow)
install_mlflow()
```

7. We will now proceed to install extra packages with the data we will need to be able to execute our example. In this particular example, we will be using the `carrier` package to facilitate the manipulation of remote functions and log information about them. We will also include the MASS package, which contains the dataset we will be using in this example. The `e1071` package and `randomforest` will be used for statistical functions and to run the prediction classifier. Here is the code you will need:

```
install.packages("carrier")
install.packages("e1071")

library(MASS)
library(caret)
library(e1071)
library(randomForest)
library(SparkR)
library(carrier)
```

8. Next, we will focus on starting the experiment by starting a block of code with this line of code: `with(mlflow_start_run(), {`. This will basically allow us to start logging the model parameters through the `mlflow_log_param` function. In the following case, we will be logging in MLflow the number of trees (`ntree`) and the number of features randomly sampled (`mtry`) at each split of the algorithm. The code is illustrated in the following snippet:

```
with(mlflow_start_run(), {

  # Set the model parameters
  ntree <- 100
  mtry <- 3
    # Log the model parameters used for this run
```

```
mlflow_log_param("ntree", ntree)
mlflow_log_param("mtry", mtry)
```

9. In the next two lines, we instantiate the `random forest` algorithm by specifying the `Pima.tr` training dataset and adding the algorithm parameters. We then predict using the `Pima.te` test data. The code is illustrated in the following snippet:

```
rf <- randomForest(type ~ ., data=Pima.tr, ntree=ntree,
mtry=mtry)
```

```
pred <- predict(rf, newdata=Pima.te[,1:7])
```

10. We can now focus on calculating metrics around model performance—in this case, specificity and sensitivity—through the `confusionMatrix` method available in the `caret` package, as follows:

```
# Define metrics to evaluate the model
cm <- confusionMatrix(pred, reference = Pima.te[,8])
sensitivity <- cm[["byClass"]]["Sensitivity"]
specificity <- cm[["byClass"]]["Specificity"]

# Log the value of the metrics
mlflow_log_metric("sensitivity", sensitivity)
mlflow_log_metric("specificity", specificity)
```

11. We can now focus on uploading a confusion matrix plot based on previous metrics. The method in R to achieve logging of the model is `mlflow_log_artifact`. Here's the code you'll need:

```
# Log the value of the metrics
  # Create and plot confusion matrix
png(filename="confusion_matrix_plot.png")
barplot(as.matrix(cm), main="Results",
      xlab="Observed", ylim=c(0,200),
col=c("green","blue"),
      legend=rownames(cm), beside=TRUE)
dev.off()
```

```
# Save the plot and log it as an artifact
mlflow_log_artifact("confusion_matrix_plot.png")
```

12. Finally, we can serialize the model function and log it in MLflow so that it can be reusable from another R notebook, by using the `crate` method available on the `carrier` package. We end up logging the model with `mlflow_log_model` and closing the code with a bracket on the last line, as illustrated in the following code snippet:

```
predictor <- crate(function(x) predict(rf,.x))
mlflow_log_model(predictor, "model")
})
```

13. You are now free to explore the **Experiment** tab on your environment, and you should have access to your model log and be able to explore the metrics and details of the run, as shown in the following screenshot:

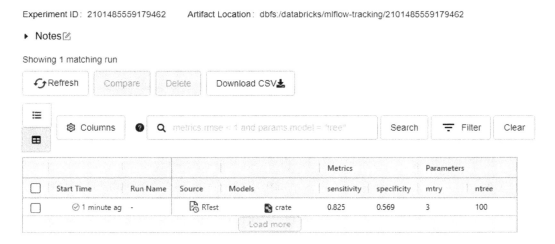

Figure 12.10 – Models automatically available in MLflow

In this section, we explored examples in Java and R, extremely relevant languages in the ML ecosystem for both engineers and data scientists. We will now delve into extending MLflow functionalities through plugins.

Understanding MLflow plugins

As an ML engineer, multiple times in your project you can reach the limits of a framework. MLflow provides an extension system through its plugin features. A plugin architecture allows the extensibility and adaptability of a software system.

MLflow allows the creation of the following types of plugins:

- **Tracking store plugins**: This type of plugin controls and tweaks the store that you use to log your experiment metrics in a specific type of data store.

- **Artifact repository**: You are able to override the artifact repositories with your own storage system—for example, adding an artifact repository based on the **Hadoop Distributed File System** (**HDFS**) or any object store specific to your environment, overriding API calls such as log_artifact and download_artifacts.

- **Running context providers**: You can update how your system logs information about the context—for instance, tags such as git_tags and repo_uri, and other relevant elements of the context of your system.

- **Model Registry store**: This feature allows you to customize where your models are stored; you can store them—for instance—in a **Secure File Transfer Protocol** (**SFTP**) system if this is the only way you might store the models of your production infrastructure. This feature can be advantageous in regulated environments where only a limited set of services and your Model Registry store need to adapt to the situation.

- **MLflow project deployment**: This type of plugin controls and tweaks how you deploy. In a case where your deployment is not for an environment supported by MLflow, you can use this feature to specialize the way you deploy.

- **Request header provider**: Enables you to control and add extra values to outgoing REST requests from MLflow. One example would be if all **HyperText Transfer Protocol** (**HTTP**) requests needed a header key related to a security token in your network that integrates with the company **single sign-on** (**SSO**).

- **Project backend**: This gives extensibility to run MLflow in different execution environments. For instance, Kubernetes is a backend as well as Sagemaker, so the integration of MLflow and the environment where models will be deployed needs specific code for each situation.

To create a plugin, you will have to create a Python package that overrides a specific module in MLflow. We will develop step by step an example MLflow plugin from the official documentation. You can follow along with the following repository URL: `https://github.com/PacktPublishing/Machine-Learning-Engineering-with-MLflow/tree/master/Chapter12/mlflow-psystock-plugin`. To run through the process, follow these next steps:

1. Define your plugin in the `setup.py` file. The `install_requires=["mlflow"]` line of code bundles MLflow with your package, being sufficient to install your new plugin package, and it will create a changed instance of MLflow. The code is illustrated in the following snippet:

```
setup(
    name="mflow-psystock-deployment-plugin",
    # Require MLflow as a dependency of the plugin, so
that plugin users can simply install
    # the plugin and then immediately use it with MLflow
    install_requires=["mlflow"],
    entry_points={
        "mlflow.deployments": " psystock target=
psystock. deployment_plugin"
    }
)
```

2. Create a package namespace empty file in a folder called `mlflow-psystock-deployment/_init_.py` to signal the creation of a package.

3. The next step involves overriding the creation of a file with methods that we want in our plugin to override the default behavior in MLflow.

   In our specific case, we will be looking at overriding the `BaseDeploymentClient` class in MLflow, which basically means that we need to implement all the methods. We will implement a set of dummy methods to illustrate the process, starting with the `create_deployment` and `update_deployment` methods, as follows:

```
import os
from mlflow.deployments import BaseDeploymentClient
p_deployment_name = "pystock"
```

```
class PluginDeploymentClient(BaseDeploymentClient):

    def create_deployment(self, name, model_uri,
flavor=None, config=None):
        if config and config.get("raiseError") == "True":
            raise RuntimeError("Error requested")
        return {"name": f_deployment_name, "flavor":
flavor}

    def delete_deployment(self, name):
        return None
    def update_deployment(self, name, model_uri=None,
flavor=None, config=None):
        return {"flavor": flavor}
```

4. We then implement the list_deployments and get_deployments methods, as follows:

```
    def list_deployments(self):
        if os.environ.get("raiseError") == "True":
            raise RuntimeError("Error requested")
        return [f_deployment_name]

    def get_deployment(self, name):
        return {"key1": "val1", "key2": "val2"}

    def predict(self, deployment_name, df):
        return "1"
def run_local(name, model_uri, flavor=None, config=None):
    print(
        "Deployed locally at the key {} using the model
from {}. ".format(name, model_uri)
        + "It's flavor is {} and config is {}".
format(flavor, config)
    )
```

The run_local(name, model_uri, flavor=None, config=None) method is the main method that will be executed upon instantiation of this plugin.

5. You can now install your plugin on top of **MLflow** by running the
    following command:

```
pip install-e .
```

We conclude the book with this section on extending MLflow with new functionalities,
allowing you as an ML engineer to extend MLflow whenever it makes sense.

Summary

In this chapter, we addressed some use cases, with example MLflow pipelines. We looked at
implementing AutoML in two different scenarios. Where we don't have targets, we will need
to use anomaly detection as an unsupervised ML technique. The use of non-Python-based
platforms was addressed, and we concluded with how to extend MLflow with plugins.

At this stage, we have addressed a good breadth and depth of topics in the area of ML
engineering using MLflow. Your next step is definitely to explore more, and leverage on
your project the techniques learned in this book.

Further reading

In order to further your knowledge, you can consult the documentation at the
following links:

*   https://pycaret.org/about
*   https://www.mlflow.org/docs/latest/plugins.html

Packt.com

Subscribe to our online digital library for full access to over 7,000 books and videos, as well as industry leading tools to help you plan your personal development and advance your career. For more information, please visit our website.

Why subscribe?

- Spend less time learning and more time coding with practical eBooks and Videos from over 4,000 industry professionals

- Improve your learning with Skill Plans built especially for you

- Get a free eBook or video every month

- Fully searchable for easy access to vital information

- Copy and paste, print, and bookmark content

Did you know that Packt offers eBook versions of every book published, with PDF and ePub files available? You can upgrade to the eBook version at packt.com and as a print book customer, you are entitled to a discount on the eBook copy. Get in touch with us at customercare@packtpub.com for more details.

At www.packt.com, you can also read a collection of free technical articles, sign up for a range of free newsletters, and receive exclusive discounts and offers on Packt books and eBooks.

Other Books You May Enjoy

If you enjoyed this book, you may be interested in these other books by Packt:

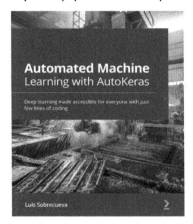

Automated Machine Learning with AutoKeras

Luis Sobrecueva

ISBN: 978-1-80056-764-1

- Set up a deep learning workstation with TensorFlow and AutoKeras
- Automate a machine learning pipeline with AutoKeras
- Create and implement image and text classifiers and regressors using AutoKeras
- Use AutoKeras to perform sentiment analysis of a text, classifying it as negative or positive
- Leverage AutoKeras to classify documents by topics
- Make the most of AutoKeras by using its most powerful extensions

Distributed Data Systems with Azure Databricks

Alan Bernardo Palacio

ISBN: 978-1-83864-721-6

- Create ETLs for big data in Azure Databricks
- Train, manage, and deploy machine learning and deep learning models
- Integrate Databricks with Azure Data Factory for extract, transform, load (ETL) pipeline creation
- Discover how to use Horovod for distributed deep learning
- Find out how to use Delta Engine to query and process data from Delta Lake
- Understand how to use Data Factory in combination with Databricks
- Use Structured Streaming in a production-like environment

Packt is searching for authors like you

If you're interested in becoming an author for Packt, please visit `authors.packtpub.com` and apply today. We have worked with thousands of developers and tech professionals, just like you, to help them share their insight with the global tech community. You can make a general application, apply for a specific hot topic that we are recruiting an author for, or submit your own idea.

Share Your Thoughts

Now you've finished *Machine Learning Engineering with MLflow*, we'd love to hear your thoughts! Scan the QR code below to go straight to the Amazon review page for this book and share your feedback or leave a review on the site that you purchased it from.

`https://packt.link/r/1-800-56079-6`

Your review is important to us and the tech community and will help us make sure we're delivering excellent quality content.

Index

www.ingramcontent.com/pod-product-compliance
Lightning Source LLC
Chambersburg PA
CBHW060542060326
40690CB00017B/3582